Praise for Designing for Wearables

"An excellent guide on how different forms of wearables developed, specific examples of what is happening now and how it works, and how people can practically frame their own ideas for product development. Use it to understand the different types of hardware and the user needs that are being met now and could be met in the future."

**—ALASTAIR SOMERVILLE
SENSORY DESIGN CONSULTANT,
ACUITY DESIGN**

"The future is here, and there's no better resource to guide us through the possibilities, promises, and pitfalls of designing for wearables than Scott Sullivan's thoughtful and timely work."

**—JESSE JAMES GARRETT
AUTHOR OF *THE ELEMENTS OF USER EXPERIENCE***

Designing for Wearables

*Effective UX for Current
and Future Devices*

Scott Sullivan

Beijing · Boston · Farnham · Sebastopol · Tokyo

Designing for Wearables

by Scott Sullivan

Printed in the United States of America.

Published by O'Reilly Media, Inc., 1005 Gravenstein Highway North, Sebastopol, CA 95472.

O'Reilly books may be purchased for educational, business, or sales promotional use. Online editions are also available for most titles (*https://oreilly.com/safari*). For more information, contact our corporate/institutional sales department: 800-998-9938 or *corporate@oreilly.com*.

Development Editor: Angela Rufino	**Indexer:** Lucie Haskins
Acquisitions Editor: Nicolas Lombardi	**Interior Designer:** David Futato
Production Editor: Colleen Cole	**Cover Designer:** Karen Montgomery
Copyeditor: Rachel Monaghan	**Illustrator:** Rebecca Demarest
Proofreader: Amanda Kersey	

December 2016: First Edition

Revision History for the First Edition
2016-12-09: First Release

See *http://www.oreilly.com/catalog/errata.csp?isbn=0636920047544* for release details.

978-1-491-94415-8

[LSI]

To Marita, who discussed in detail every paragraph in this book for an entire year, and still stayed with me.

[*contents*]

[*Foreword*]

WEARABLES IS A BROAD STROKE. A word that encompasses so many different things and contains a lot of underlying meaning and history. On the one hand, it implies a technology that we can wear—one that is close to the body, intimate, and personal. On the other hand, the things we wear can also be fashion, something that we use to express an outward appearance, convey personality, belonging, allegiances, and our existence in a specific place and time. As Scott so astutely explores in this book, when at their best, wearable tech encompasses both of these things and more. At their worst, they do neither and are a fleeting novelty.

We've been wearing technology for many thousands of years, everything from insulation to keep us warm, to mechanical watches that track the time. So, what's new about the current branch of wearable tech? How does it bring new and exciting possibilities to life, and what is our role as designers in this world? That is the topic of this book, and one that is explored and elaborated on in depth in the next 11 chapters.

It's an exciting time to be a designer working in technology. Miniaturization, toughness, price decreases, and long battery life have moved computing into new places and packages. Our phones and personal devices are now substantially faster and more capable than the super computers I used at university, and they're loaded with purpose-built sensors that collect data on everything we do. With this capability comes a lot of opportunity for new creations, both good and bad. As we move computers closer to our bodies and imbue them with the power to understand our physicality and psychology, we need to think carefully about how we use these powers. When the devices we use affect the way we understand ourselves through our bodies, fitness, diet, and more, the interactions can be fraught with psychological and social nuance. The way we design these interactions can have a huge impact on the lives of the people who use them, helping them or reinforcing negative aspects of culture. Scott does a great job of exploring these potential pitfalls in this book, and along the way he talks about some of the history that got us here.

Fitness trackers and the like are the most common today, but new forms of wearable tech are emerging. *Sousveillance*, a term coined by the godfather of wearable computing, Steve Mann, is the practice of recording from within an activity, as opposed to surveillance, which is recording something from above, or outside, of the activity itself. Wearable always-on cameras let people record, *sousveil*, their life as they go through it. Mann has been wearing a camera and computer every day since the 1980s, capturing his life from his own perspective. As this tech becomes cheaper and smaller, more and more people are incorporating it into their daily wear. What does it mean to be watching and recording one another all the time? How does it change our relationships and public social activity? These are interesting questions that should be explored through a reflexive practice of prototyping and research.

Over the following pages Scott Sullivan elaborates on these ideas through examples, how-to, and history. He weaves together many disparate ideas that show us how wearable technology is more than just the device that sits on your wrist/neck/clothes; it is also the invisible services, culture, and behavior behind the devices that give them their power. Ultimately, wearables aren't about putting computers on your body; they're about changing the way we live and behave together. They can change the way we understand one another and how we present a picture of ourselves through outward appearance augmented with data.

This book is a beginning. A thorough and enjoyable beginning that gives us some history, some philosophy, some practice and technique, and a foundation upon which we can build a deep and thoughtful approach to designing wearables. I can't wait to see what we do with the information and techniques that follow.

—MATT NISH-LAPIDUS,
INTERACTION DESIGNER AND
CREATIVE TECHNOLOGIST

[*Preface*]

Because We Can

We're at a crossroads right now in terms of wearable devices. On the one hand, we're still exploring: new technologies are pushing the envelopes of both technical ability and social conventions, reshaping how we understand ourselves and the world around us. On the other hand, we're beginning to see patterns and we're seeing the commoditization of certain ideas that have been introduced in only the past couple years. This is a tipping point at which we can look back at products that we've been living with, understand the impact of those products in the greater ecosystem of our lives, and reflect on whether those contributions are worth keeping around or whether we are better off doing something else.

I think of this moment as being a wave that we're surfing. The wave is propelled by the idea of "because we can," meaning that we figure out how to do something from a technological perspective and introduce a new product to the public as a hypothesis. Because the technology is so new, we really don't know whether the product will be beneficial or useful, because it's never existed before. A lot of the time, technologies will be released into the world with a specific use case and prove to not only fail to deliver on the use case, but also to compound the issues, real or invented, that they're claiming to alleviate. For example, a smartwatch that is touting the ability to simplify your life constantly commands your attention and pulls you out of situations that you'd otherwise be focused on, or a fitness tracker that claims to give a full picture of your health, but only has the ability to tell half of the story.

WHAT IS DESIGN'S ROLE IN THIS?

I'm sure when most people think about designing a wearable device, industrial design comes to mind. After all, we're talking about physical devices and the most obvious designed element is the device itself. If we want to talk about user experience (UX) design specifically, we can talk about the interface of the device, like the screen on a smartwatch, or any number of configurations of LEDs on a fitness tracker, because most popular definitions of UX tend to skew toward digital interfaces, or at least manifest as them. The specific design role that I feel is most appropriate here is *service design*. Here is how the Service Design Network (SDN) defines service design:[1]

> the activity of planning and organizing people, infrastructure, communication and material components of a service in order to improve its quality and the interaction between service provider and customers.

Thinking of design for wearables as service design works best because it's not encumbered by the silos of the design process, having access to the product as a whole instead of one aspect of the design. There are a lot of decisions and experiences that happen in those gaps and service design can be accountable.

Thinking of wearables as a service also speaks to the nature of the devices themselves. The value of any given wearable device does not come from the device itself; rather, it comes from the service that it provides. Mike Kuniavsky refers to the devices as *service avatars*, because the devices are generally single-purpose, low-power objects that act only as a gateway to their cloud-based accompanying services.[2] For example, let's look at the functional difference between a pedometer from the 1990s and any Fitbit; the devices themselves measure the exact same thing, but the reason pedometers didn't balloon in to a multibillion-dollar industry in the 1990s is that they provide no additional insight into the data that they measure.

1　Meghan Lazier, "What Is Service Design?" *Designlab Blog*, Accessed October 1, 2016 (*http:// trydesignlab.com/blog/what-is-service-design*).

2　Mike Kuniavsky, "8.2 Devices Are Service Avatars," in *Smart Things: Ubiquitous Computing User Experience Design* (Amsterdam: Morgan Kaufmann Publisher, 2010), 102–104.

Who Should Read This Book

This book *is* about the design of these services and the devices that support them, but it's about a lot of other aspects of wearables, as well, such as sales, adoption of new technology, the history of wearables, and research.

This book should be particularly useful to you if you:

- Design and/or develop software

- Design and/or develop hardware

- Sell or attempt to sell software or hardware

- Want to launch a wearable product

- Work in the Internet of Things (IoT)

- Are at all interested in wearable technology

- Have given up on wearable technology

How This Book Is Organized

This book comprises 11 chapters. Here's what each chapter discusses:

Chapter 1: Design Follows Technology
> This chapter focuses on the history of wearable devices and the particular decisions that have shaped how we think about wearables.

Chapters 2 through 8 go through individual types of wearables, beginning with the most common available as of this writing, and ending with the emerging and future-focused categories. Each chapter focuses on experience from embodied research as well as interviews of people about their experiences.

Chapter 2: Activity Trackers
> Chapter 2 covers the evolution of fitness trackers beginning with early mechanical devices. We will cover how the devices are sold, how they're perceived in society, where they excel and how they fall short of providing a complete health picture.

Chapter 3: Smartwatches

In this chapter, we cover the evolution of watches that do more than just tell time. We discuss the long history of watches with exceptional functionality, how they currently work in our technological ecosystem, and how they fit in to our lives.

Chapter 4: The Glass Experiment

Chapter 4 looks at the rise and fall of Google Glass as a consumer wearable device. We talk about its development, launch, public perception, and the reasons why the product eventually failed.

Chapter 5: Wearable Cameras

Chapter 5 begins with discussion of currently available wearable devices such as the Narrative Clip and GoPro cameras. It then moves on to a discussion of design considerations when dealing with high volumes of image data, and then explores what services are possible with this information.

Chapter 6: Cognitive Wearables

This chapter covers the growing product category of *cognitive wearables*—wearable devices that measure, affect, and optimize our cognition. There are a number of issues with both objects based on the seemingly subjective nature of cognition itself and the level of insight they provide.

Chapter 7: Service Design

Here, we provide an introduction to designing wearables out of the "Wild West" phase and into the world of truly valuable and considerate services. The difference from being a cheerleader—"Only 3,000 more steps"—to a coach—"You're getting less sleep, and that's having an effect on your overall health in *this* way, and you need to do *this*." The reach and tools of the field of service design will be what turns wearable devices from novelties to legitimate tools.

Chapter 8: Embodiment and Perception

Chapter 8 covers two key concepts about how these devices/services are designed. We discuss how our physicality shapes our understanding of ourselves and the world around us, and how certain products have an effect on that physicality, and thus, our behavior.

Chapter 9: Prototyping

Chapter 9 looks at current prototyping processes and tools such as Arduino and Processing. This is an overview of the tool capabilities, and then how to use these tools to prototype and validate ideas, which isn't all that much different from the screen-based prototyping that you're most likely used to.

Chapter 10: Selling the Invisible

This chapter covers the very difficult task of introducing new technology to a market that has little to no context for what the device does or the potential impact of the service that it enables. We discuss the importance of storytelling and the innovative ways retailers are introducing people to technology.

Chapter 11: Moving Forward

Finally, we end by taking a look at where wearable technology is headed and how this technology will change our relationship with computers as a whole.

Acknowledgments

First, I'd like to thank my wife, Marita, who gracefully endured a year of her husband being absent for most nights and weekends while writing this book. Marita patiently helped me work through some of the more complex concepts of this book and was always available to give advice and feedback. There's no way I could have accomplished this without her.

I'd like to thank Angela Rufino for being an amazing editor and being very patient with me through the very long process of writing this book!

Thanks to Erik Dahl who taught me how to be a designer; that design cannot be confined to traditional media, processes, and artifacts; and that the role of the designer is fluid. Many of the foundational concepts of this book are a direct result of explorations with Erik, and I'm very lucky to continue to work with him.

Thanks to everyone at Adaptive Path and Capital One, from whom I have learned so much from over the years. Thank you, Maria Cordell, for being supportive of my technological forays. Thank you, Brandon Schauer, for the writing feedback. Thank you, Jesse James Garrett, for listening to all of my ideas and clarifying everything. Thank you, Nick Remis and Jamin Hegeman, for helping me talk about service design without sounding like an idiot. Thank you, Peter Merholz and Chris Risdon, for the many discussions that shaped my understanding of wearable technology. Thank you, Kristin Skinner and Patrick Quattlebaum, for the writing advice and support.

Thanks to Alastair Somerville for letting me bug him with questions for hours on end and providing feedback on the book. Thank you, Matt Nish-Lapidus, for many of the foundational concepts of this book, amazing feedback throughout the way, and for writing the forward. Thank you, John Follett, for support throughout writing this book, the great feedback, and the writing advice in general.

Thanks, Tayler Blake, for teaching me about data science and providing a ton of feedback on the book.

Thank you, Frank LaPrade, for reminding me about how mind-blowingly cool this technology has always been and for the support while writing the book.

Thank you to Andrew Hinton, Mike Kuniavsky, Alex (Sandy) Pentland, Haig Armen, and Karl Fast for being major sources of information and inspiration for this book.

Thank you, Kelly Beacham and Colby Beacham, for continued feedback on a bunch of different devices and concepts. And, finally,

Thank you, Frank Sullivan, for sitting next to me while writing the entire book and being an amazing listener.

Design Follows Technology

With all disciplines of design, as technology progresses, the restrictions become fewer and fewer until they get to the point at which a wave crashes and there's nothing holding you back from making anything you can dream of. Architecture became bolder when steel frames could support nearly any structure architects could dream of without being restricted by the need to cater to gravity to the degree they needed to when building with stone alone. Furniture evolved with advancements in the technology of the materials available to designers, such as fiberglass and bending plywood, as well as technological advancements related to mass manufacturing processes. In graphic design, typefaces became increasingly more complex and detailed as printing technologies evolved to the point that those details could be reproduced accurately, and then it went completely insane when computers removed nearly all graphical restrictions.

Digital Maturity

Wearable technology has been around in one form or another since 17th-century China when someone came up with an abacus that you could wear on your finger. So, what's so special about wearable technology at this point in time? There's a reason wearable technology is making a resurgence. It's because we've finally made it to the digital maturity tipping point: the point at which advancements in computing and the supporting infrastructure have lessened the restrictions of the design of technology so that they're nearly nonexistent.

Considering that most people use their computers for simple internet-based information recording and retrieval, the gap in capabilities of a laptop and a computer the size of a watch is not very significant, at least from a technical standpoint. Though physical size of the technology is a big part of this advancement, there's also the cost of computing, which has fallen to such a degree that this type of technology can proliferate on a major scale. Combine this with the open source hardware movement that is making the tools to create this technology available to a wider range of people. In this section, we go over some of the specific things that happened to advance wearable technology to where it is today.

INTEGRATED CIRCUITS

On a high level, processors work in pretty much the same way they did in the 1950s, when the first silicon chip was invented: a series of switches called transistors that have an on state and an off state control the flow of electrons. These are the 1s (on state) and 0s (off state) that compile into what we see on our computer screens. In 1965, Gordon Moore wrote an article for *Electronics* magazine in which he predicted that the number of transistors that could fit on to a chip would double every two years.[1] This became known as *Moore's law* (Figure 1-1). Currently, the smallest transistor is down to 14 billionths of a meter wide. Beyond size efficiency, integrated circuits have seen increases in energy efficiency on par with Moore's law, which is also very important for computers that can't be plugged in to a power source all the time.

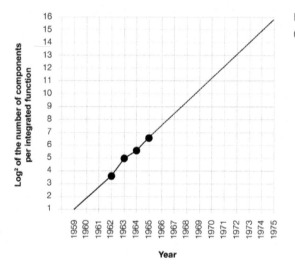

FIGURE 1-1

Graph of Moore's law

---

1 Gordon E. Moore, "Cramming More Components onto Integrated Circuits," *Electronics*, April 19, 1965.

THE INTERNET

The internet is obviously a major contributor to the advancement of wearable technology: the data that's retrieved and recorded on a wearable device doesn't need to be stored on the device if it can be stored on a server. The internet is also making Moore's law slightly less relevant as a computational speed limit because said computation doesn't need to be done entirely on a local device. An example of this are image recognition programs on Google Glass that take a photograph and immediately upload it to a server for processing instead of trying to do the computation on the device itself.

CELLULAR DATA NETWORKS

Cellular data is the primary access point for untethered wearable devices, whether through a phone or the device itself. 4G LTE cellular data has a maximum speed of 100 megabits per second (MBps) for downloads and 50 MBps for uploads, which is fast enough for most wearable applications.

OPEN SOURCE

The last major contributor to the perfect storm of wearable computing is the increased level of access to the tools used to create the technology itself. This is primarily due to the open source movement, wherein the source code of software and original designs of devices are made available to the public to reproduce and modify freely. In addition to the designs and technology itself, an incredibly supportive educational community has grown around the open source movement.

The Human Problem

So, we're now free of limitations with respect to the size of the device, the communication infrastructure, and even access to the production of the technology, but wearable devices are limited by one major thing: humans. Now that we're free to do whatever we want with technology, we need to figure out how we actually want to move forward. The two main human-related areas are input—how we put information into the computer—and output—how the computer gives information to us.

INPUT

There are two high-level types of input for wearable devices. First, there's detailed input such as text input or manipulation of objects and menu systems, basically the type of input you'd need to operate a normal desktop computer. Then, there's passive input, which refers to things like steps, heart rate, GPS location, ambient noise, and other things that can be collected passively.

Detailed input

Detailed input on wearable devices is difficult, primarily because of the prevalence of the keyboard and mouse. People don't necessarily want to keep using the keyboard and mouse as primary input mechanisms; it's just that they're so dominant that the way we design and think about digital interactions are inevitably stuck in that paradigm. It's similar to television in its earliest days, when there just wasn't an established content paradigm for the new medium, so most of the early television shows were just people reading radio scripts. Eventually we'll get to something more appropriate, but it will take a while to turn the ship.

Today, there are a few techniques being tried out in terms of wearable text input. One-handed keyboards (Figure 1-2) were popular in academic circles but are generally considered clunky and difficult to learn. Voice input is getting a lot stronger in terms of recognition, but inevitably comes with some issues such as other people being able to hear what you're saying and background noise issues. The cleanest method of text input that I've seen is actually instructing users to pull out their phones and use them as a temporary keyboard.

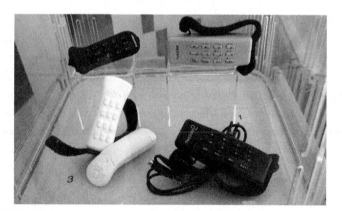

FIGURE 1-2
Twiddler one-handed keyboards

Beyond text input, there has been some interesting progress in gestural interfaces, particularly with the Myo Armband (Figure 1-3) from Thalmic Labs, which reads EEG signals from the movement in your arm muscles. Microsoft's Hololens (Figure 1-4) also uses gestures as a primary input mechanism by targeting an object by looking at it and gesturing toward that object. In addition to gestures, Microsoft released a "clicker" which is essentially a handheld mouse trackpad to use with the Hololens.

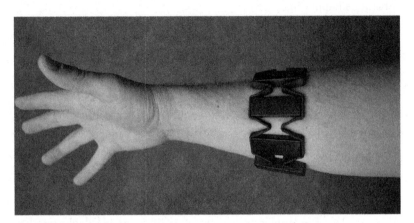

FIGURE 1-3
Myo armband

FIGURE 1-4
Microsoft Hololens with clicker

Passive input

Passive input is a lot easier to figure out from a design perspective. It's basically a matter of finding indicators of certain behaviors; for example, if the accelerometer on my device is moving in a specific pattern, I can safely assume that it's an indicator that I'm walking and each cycle of that pattern is a step.

The most basic example of passive input are fitness trackers. These devices are worn all the time and you don't really interact all that much with them other than swinging your arms around. A more interesting example of passive input comes from MIT Media Lab spinoff Humanyze, which distributes sensor-laden lanyards to the employees of a company it's working with to analyze where and how social interactions are happening within the company's building by tracking the employees' location and recording who is speaking when there are conversations happening.

OUTPUT

Output is a much easier problem to tackle with wearable devices. Output is divided into active and passive categories. Active output includes things like heads-up displays, screen-based smartwatches, and audio feedback. These are all things that get your attention and inform you of something. Passive output is information that's displayed more like dashboards, which are typically self-initiated or even communicate without commanding active attention.

Active output

We see a lot of examples of active output in wearable devices today. These are the types of interactions that require a conscious engagement, such as when you get an email notification on your smartwatch, or a calendar reminder pops up on your Google Glass. It's pretty fair to say that most notifications are active outputs, but other non-notification interfaces, such as the display on a Fitbit, are active in the sense that in order to use them, you must actively engage with the device.

Passive output

Passive output is very different from active. Passive outputs are interfaces that might be always on and provide information to us without commanding our full attention. A great example of this is the Lumo Lift (Figure 1-5), a "digital posture coach" that vibrates every time you slouch. After a couple hours using the Lift, you get used to it to the point where it's almost like another sense—you're out and about doing things and working on things and you feel that little vibration and you react to it without diverting your attention. Another example of this would be the Withings Activité watches (Figure 1-6) that have a completely passive, always-on interface that you can glimpse out of the corner of your eye without actively thinking about.

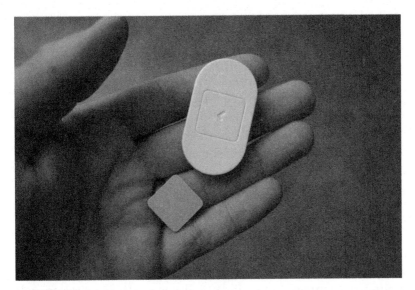

FIGURE 1-5
Lumo Lift digital posture coach

FIGURE 1-6
Withings Activité Steel

The "Us" Part of Maturity

The other side of the digital maturity coin is that now, with little to no technological restrictions on design, we must decide how we, as a society, want to live with technology. Now that we can do whatever we want, we must determine what it is *that* we want. What's healthy? What actually improves our lives? This is the part for which human-centered design is most important. We've made it through the Wild West days during which we had no idea if something was going to be useful or awful. Now, we need to take a firm look at what exactly is going on, if we're actually benefiting from technology, and why we're using that technology.

Moving forward in the design of wearable technologies requires us to be fully aware of the behavior that these devices encourage, good and bad, to provide truly useful services that empower people with more control and awareness in their day-to-day lives. We now have enough information to step back and take an honest look at how people actually use this technology so we can be critical of what we're putting into the world and be fully aware of the ramifications of introducing these products into people's lives.

Activity Trackers

In this chapter, we take a look at fitness trackers like Fitbit, MisFit, and Jawbone UP, beginning with the early trackers and what they gave us, to the more contemporary ones that employ machine learning to provide an ongoing service. Here, we'll more clearly draw the lines around services as opposed to commodities and look at how people understand tools as opposed to service delivery. We also take a quick look at some of the technology that powers these types of devices, how some of that technology is now commoditized, and how certain services can fight technological commoditization.

Early Step Tracking

The concept of the pedometer is largely credited to Leonardo da Vinci, who built a number of distance-measuring devices to aid in his cartography. In one of his sketches he added a pendulum to an early version of a surveyors wheel in order to explicitly track the number of steps taken by Roman soldiers (Figure 2-1). Pedometers became somewhat popular with European watchmakers in the 1700s and were of particular interest to Thomas Jefferson and James Madison, both of whom documented their interest in various letters. In one of those letters, Jefferson describes a pedometer watch he found in Paris in 1786:

> It has a second hand, but no repeating, no day of the month, nor other useless thing to impede and injure the movements which are necessary. For 12 louis more you can have in the same cover, but on the back side & absolutely unconnected with the movements of the watch, a pedometer which shall render you an exact account of the distances you walk. Your pleasure hereon shall be awaited.

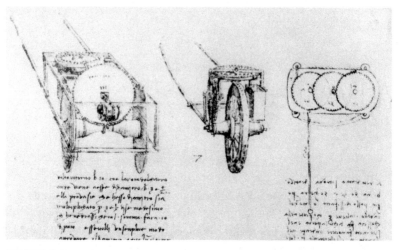

FIGURE 2-1
Sketch of Da Vinci's pedometer

Pedometers became somewhat popular in the United States in the mid-1900s. In 1938, they were marketed as toys, and the Hike-O-Meter was a premium prize you could order from Wheaties cereal (Figure 2-2). Similar aluminum pedometers were sold in the United States throughout the 20th century, but they didn't attain widespread use until the early 1960s, in Japan. In the early 1960s, Dr. Yoshiro Hatano was doing research on the walking activities of Japanese people and found that the average person walked about 3,500 to 5,000 steps per day, but in order to be healthy, they should walk 10,000 steps each day to burn 20% of their caloric intake. In 1960, Dr. Hatano started selling a pedometer called the Manpo-kei (Figure 2-3) which translates to "10,000 steps meter" and it was hugely successful. It's estimated that even today Japanese households average 3.1 pedometers. Digital pedometers started popping up in the mid-1980s in Japan and slowly made their way to America shortly after.[1] In the 1990s, they were somewhat common and even began to show up on (Japanese-made) digital watches.

---

1 "Why 10,000 Steps and Not 14,323 Steps?" *Pedometer Reviews - Each Step You Take*, last modified April 29, 2014 (*http://eachstepyoutake.com/why-10000-steps*).

FIGURE 2-2
Wheaties Hike-O-Meter

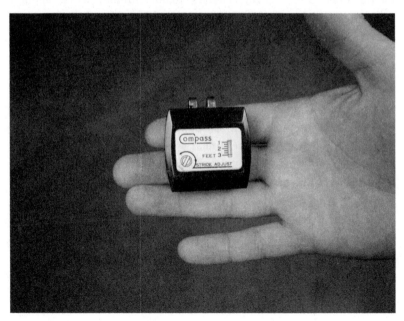

FIGURE 2-3
Manpo-Meter

Connected Fitness Trackers

As popular as early digital watch pedometers were, they lacked one very important function, data logging. If you wanted to have any awareness of your progress over time, you had to write it down somewhere, which greatly limited the usefulness of the technology. However, in 2006 Nike introduced the Nike+iPod Sports Kit, which included a sensor that was built to fit into certain Nike gym shoes and wirelessly connected to iPods or iPhones to log your data. The Nike+ kit wasn't meant to track your steps all day; it was only intended for working out, but the logging functionality turned out to be crucial to the adoption of the technology and ended up leading to more widespread production and use of fitness trackers.

CONNECTING THE DOTS

Connected fitness trackers really began to pick up steam in 2012 with the introduction of the Fitbit One (Figure 2-4), a fob-looking tracker that users clipped on to their pants, bra, or shirt. The default step goal for the Fitbit One was Dr. Hatano's 10,000 number. It had about a 10-day battery life and tracked steps, flights of stairs, and sleep (it came with a sleeve that you could put it in for sleep tracking). The Fitbit One was my first fitness tracker; everybody in my studio got them, and I was really excited about it because this was the first time I was going to be able to break that barrier between the physical and digital world! I was used to being able to access this type of behavioral information in the more conventional (screen-based) products I worked on, and it had been incredibly useful. At the time, I was working on a big project for Scholastic that would tie together a bunch of data to target reading levels of students and push them to pick new books that would help them grow as a reader. Thus, I was understandably enthralled with the possibilities that encoding physical data represented.

FIGURE 2-4
Fitbit One

In reality the physical barrier crossover wasn't the tectonic shift that I had been hoping for: the data I could get from the tracker was aggregate and limited, but it did give far more insight into the black box of physical activity than we had before. Even at this low fidelity, activity and other metrics were interesting to see. As it turns out, I was pretty sedentary, and I really had no idea, I immediately began walking more and taking the steps instead of the elevator at work. In addition to the basic input-and-output type correlations (I feel better if I walk more than 11,000 steps), one of the more interesting outcomes of continued use of these devices is knowing what certain things feel like in the context of your life. When you walk 10,000 steps for the first, second, third, twentieth time, and so on, you begin to get an idea as to what that day looks like as a whole within the very specific context of your life. You understand what changes you need to make to your lifestyle to achieve 10,000 steps per day. This is *experiential knowledge,* and it's different from *explicit knowledge* because it must be experienced. It's also different from *tacit knowledge* because it's not universal (like riding a bike). It really hinges on combining the repeated tacit experience of walking 10,000 steps and combining that information with the observation of lifestyle changes to your personal situation to take those 10,000 steps.

In the case of my day, I knew that I had to walk to work, take a little walk at lunch, and walk my dog for a little longer than usual if I wanted to hit my 10,000 step goal every day. For a lot of people, gaining this experiential knowledge—which took maybe a week to achieve—was the primary benefit of owning one of these early fitness trackers, and once that information is more encoded into your lifestyle and daily routine, the usefulness of the tracker begins to decline. When you know what you need to do to achieve this goal and you do it repeatedly, the specific measurements of this information become less and less useful to you. As the device provides less and less value, you get to a point at which the benefits outweigh the costs of upkeep. With the Fitbit One, upkeep meant charging it every 10 days or so, and always remembering to take it out of your pants when you were doing the laundry and clipping it to yourself every morning. For multiple members of my studio and other friends who bought the device, this was all they needed and they slowly began falling out of rotation after one or two charge cycles. For the people who did stick with the device, the emphasis switched over to the more social aspects like competing with friends.

Not All Steps Are Equal

After the commercial success of the Fitbit One, there was a flood of new fitness trackers. One big lesson everyone seemed to learn was that the pager-style form factor of the Fitbit One was inconvenient, and so it was decided to move the pedometers to the wrist, which might be the worst place on your body to measure your steps. The primary emphasis of steps, (specifically, Hatano's 10,000 steps goal) never went away. A major highlight of this first group of devices was Nike's FuelBand (Figure 2-5). The FuelBand was aimed at more athletic individuals and still tracked the dubious life-based metrics like steps and an estimated calorie burn, but the interesting part of the FuelBand ecosystem was how much emphasis that Nike placed on Fuel Points (Figure 2-6), a proprietary unit of measurement that attempted to describe overall energy expenditure. Nike's Fuel Point system was an attempt to even out the inherent flaws of using the step count as the metric that activity goals are based on, as opposed to using step count and then displaying calorie burn as a secondary metric.

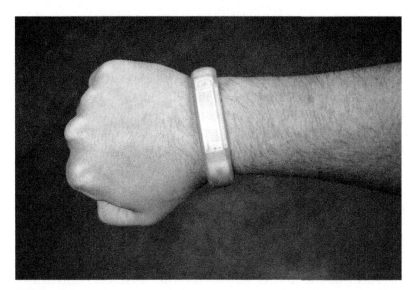

FIGURE 2-5
Nike Fuelband

FIGURE 2-6
Nike FuelBand application screenshots

The problem with basing your activity goal on overall steps is that all steps are not created equal. 10,000 steps for someone who weighs 150 pounds is very different in terms of energy expenditure than it is for someone who weighs 250 pounds, as is 100 steps down hill as opposed to 100 steps up a steep incline, or the difference between running as opposed to casual walking. These are all factors that overall steps don't account for, but there's also the issue of vigorous athletic activity that must be measured by an accelerometer and gyroscope that's worn on your wrist. I would not want to be on the data team that was tasked with translating the complex movements of someone playing basketball into an accurate step count. Even if they perfected those algorithms, what use would the output be when a leisurely hour-long stroll would produce the same measurement? Most trackers picked up this slack by estimating how many calories one burned, but the difference between the FuelBand and everyone else was that Nike used Fuel Points to integrate energy expenditure into the primary metric that people used to measure their activity.

More of the Same

Each new device in this first group had its own small advancements or distinctions. The Misfit Shine was appealing because it used a watch battery that lasted forever and didn't look like a rubber band. Fitbit Flex was like the Fitbit One except it was worn on the wrist; the Jawbone UP24 band was designed by Yves Behar and had a better designed interface; and the Withings Activité (Figure 2-7) was really cool because it looked like a normal watch with an analog interface. All of these devices were slight improvements layered on top of the same underlying functionality, a gyroscope and accelerometer that helped you count to 10,000 steps. The canary in the 10,000-step coal mine came from the Chinese firm Xiaomi when it released its flagship Mi Band (Figure 2-8). The Mi Band fits in with this group really well: it tracks steps and sleep with surprising accuracy, lets you compete against friends, has a decent battery life, syncs with a phone, and even vibrates when getting a call. The reason the Mi Band is interesting is because you can get one for about $10 where the other devices in this category cost at least eight times as much. The fact of the matter is that the sensors and underlying reporting have become so cheap that to stay in business, you need to go beyond reporting of basic activity.

FIGURE 2-7
Withings Activité Pop

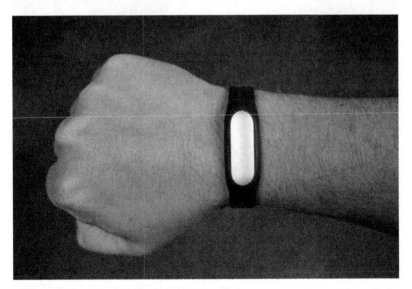

FIGURE 2-8
Xiaomi Mi Band

More Sensors and Machine Learning

Now that the new reality of the fitness tracker market has been accepted, there is a new generation which is centered around heart rate. The two leading devices of this new era are the Fitbit Charge HR (Figure 2-9) and the Jawbone UP3. The Fitbit uses a photoplethysmogram (PPG) sensor to measure your heart activity. This sensor tracks optical changes in the light absorption of the skin, which indicates heart activity. Jawbone's UP3 uses bioimpedance, which sends a small amount of electricity through the surface layer of the skin and gleans information through changes in the resistance to that electric signal. Though these additional sensors are what's marketed as the primary selling points of these advanced activity trackers, what will define this generation will come from something else altogether—*machine learning.*

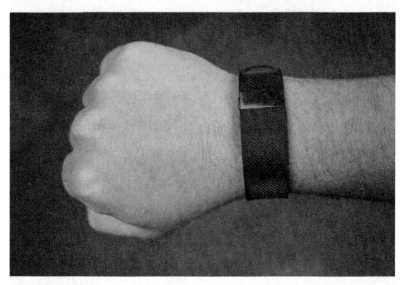

FIGURE 2-9
Fitbit Charge HR

Before going into the details of the algorithmic capabilities of these devices, I want to talk about the difference that application of data makes in the context of providing a service. In older economic terms, services are things like education, and commodities tangible items like salt. A popular saying among my more service-oriented friends is, "If you can't drop it on your foot, it's a service." Both the Fitbit Charge HR and the Jawbone UP3 fall in the spectrum of commodity to service, but they're at opposite ends of this spectrum. The Fitbit uses a "cheerleader" model, in which you'll get periodic updates (mainly on the device itself) that amount to a contextual check-in—something like, "only 3,000 more steps to go!"—and in this way, the Charge HR isn't all that different from the 10,000-step generation of trackers.

The Jawbone UP3 (Figure 2-10) stands out from nearly every other tracker on the market by positioning itself as a service that is powered by machine learning. Compared to the Fitbit's cheerleader model, the UP3's service is a "coach" model in that it records your information, makes sense of it, and, based on this information, will suggest a change in behavior. Instead of "Only this many more steps," the UP3 will say, "You did this, it means this, and you should do this." Internally Jawbone refers to this as the "Track, Understand, and Act" model,[2] in which, through machine learning, the service can make certain assumptions about your activity, and then via its prescriptive, it can tell you what to do based on the understandings of your activity. And that's where the value comes from: it's now a service.

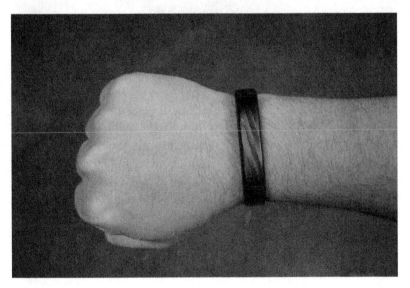

FIGURE 2-10
Jawbone UP3

2 "Jawbone Introduces UP3," Jawbone Press Release, November 4, 2014 (*http://content. jawbone.com/static/www/pdf/press-releases/jawbone-introduces-up3.pdf*).

Figure 2-11 is an example of how the Track, Understand, and Act model might work. After a night of heavy drinking, the three bioindicators that Jawbone could use to understand my behavior are:

- I was physically active later into the night than I usually am.

- My RHR was higher when I was sleeping.

- My sleep was a lot more restless.

FIGURE 2-11
Jawbone Smart Coach screen

The understand part of the process then concludes, "Scott was drinking last night," though it doesn't expose the understand part to me. The only part that I'm actively aware of is the prescription, which is that I should drink a lot of water and go to sleep earlier. When I'm not drinking, a more typical message might compare my steps to my average, or let me know how different aspects of my behavior are related, such as, "You were awake about 38 minutes earlier than usual this morning. You tend to move less after an early rise. For you, 38 minutes means about 542 fewer steps. Start a new trend. Head out and maintain your 12,731 step average."

Another thing that is interesting about the UP3 is what it can do, yet doesn't thus far. If you look at the box of the Jawbone UP3, there are three things in the specifications section that aren't activated in the software at this time: a heart flux sensor, skin temperature sensor, and galvanic skin response. These things add up to what could basically be used as a rudimentary polygraph test and place the UP3 in the realm of

cognitive wearables, which we discuss in Chapter 7. What is most exciting about this scenario is how the cognitive data could be combined
with activity data to provide more useful information such as, "Your
step average is down by about 700 steps this week. For you, this means
about a 28% increase in overall anxiety."

Telling the Rest of the Health Story

This might sound like a weird statement, but one of the biggest problems with most fitness trackers is actually their emphasis on steps.
There are tons of studies on the benefits of walking more—I am in no
way saying that it doesn't contribute to overall health, and it's definitely
not a bad thing to be doing—but if we take a step back and look at the
primary reasons people use fitness trackers, how big of a factor in overall health and weight loss are step counts? According to the National
Institutes of Health, about a third of adults in the United States are
obese, and according to the Center for Disease Control (CDC), the leading cause of death in the United States is heart disease, which is linked
to obesity as well as other causes.[3] Addressing weight and obesity problems comes down to a simple equation known as *energy balance*, which
refers to the relationship between calories consumed over time and calories burned over time; if you consume more than you burn you gain
weight, and if you burn more than you consume, you lose weight.

The best illustration of the relationship of diet and exercise to weight
control comes from Aaron E. Carroll of the Indiana University School
of Medicine. Carroll states that exercise consumes far fewer calories
than people think. To drive this point home, he compares an exercise
regimen of 30 minutes of jogging or swimming every day to the equivalent in calorie intake reduction of eliminating two 16-ounce sodas a
day,[4] which is a much more realistic lifestyle change for most people.
With this in mind, fitness trackers are only telling half the story of our
overall health. The emphasis on physical activity is at best incomplete,
but it also can be leading people to believe that they're doing more to

---

3 "Leading Causes of Death," *CDC/National Center for Health Statistics*, last updated
   October 7, 2016 (*http://www.cdc.gov/nchs/fastats/leading-causes-of-death.htm*).

4 Aaron E Carroll, "To Lose Weight, Eating Less Is Far More Important Than Exercising
   More," *New York Times*, June 15, 2015 (*http://www.nytimes.com/2015/06/16/upshot/to-
   lose-weight-eating-less-is-far-more-important-than-exercising-more.html?_r=0*).

affect their health than they actually are, which can result in lack of noticeable progress and ultimately loss of interest in the device. So, the big question is, what can we do to fix this?

There are a few ways in which people try to track calories. The most popular is manual logging using an application like MyFitnessPal, which has a database of common foods that you can select from and makes calorie tracking relatively easy and integrates with a lot of activity tracker software. The problem with manual logging, though, is that it's not very sustainable. Compared to the passive fitness trackers that we're used to wearing, manual logging requires a lot of upkeep because it requires you to pull out your phone and punch in data after every meal. Personally, I've never been able to actively log food data for longer than a month, and I've heard the same thing from many people, even after seeing positive results.

I've always thought of passive calorie counting as a holy grail of wearables and something that wouldn't happen for a while. I remember sketching out fantasy ideas for sensors that could replace our molar teeth that would be able to monitor our food consumption or wearable cameras that would identify the food we were eating and automatically log it. It wasn't until recently that I decided to give the controversial GoBe device (Figure 2-12) a shot. GoBe had a very rocky start: the device's manufacturer claimed it could passively and noninvasively measure calorie intake with its wrist-based device, which launched in 2014 on the crowdfunding website Indiegogo, but almost immediately after raising more than a million dollars, critics began to chime in. The criticism was led by PandoDaily's James Robinson, who wrote extensively about the dubious claims of the device. But what really stood out to me was when medical professionals began weighing in. The big quote for me was from Michelle MacDonald, a clinical dietician at the National Jewish Health hospital in Denver, who said that the technology is plausible, but we're not there yet, "but when it does it will be the size of a shoebox...It will come from a big lab, will be huge news, and make a lot of money." This was enough for me and a lot of other people to completely write off the product, I didn't think about it again for almost two years.

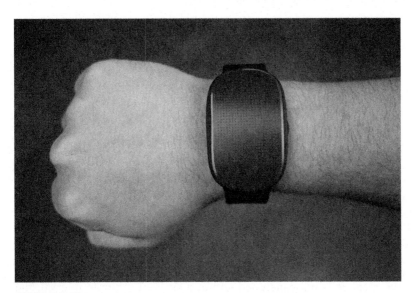

FIGURE 2-12
Jawbone Smart Coach screen

I was running some concepts for this chapter by someone who works in the wearables industry, and when I mentioned passive calorie tracking as a pipe dream, he told me to take another look at the GoBe and mentioned that researchers at his company have confirmed that it actually works. Fast forward a week or so. I have it on my wrist, it's massive and clunky looking, and it takes forever to set up on the application. But after about an hour, I had it monitoring my food intake. Here's how the mostly automatic calorie tracking interface works: a little while after your body begins metabolizing the food, you open the application and are greeted by a screen that asks you when you ate. The application knows you ate something within a certain time span, and you confirm the more specific time by turning on or off switches marked with 15-minute intervals. This is the first big communication issue: the application immediately gives me a reading of my calories when my body has only begun to metabolize the meal, so if I know for a fact that I had a 400-calorie breakfast, but a half-hour after I eat I see that the GoBe only logged it as 100 calories, it seems wrong to me. Holding off on displaying calorie counts until the meal is digested, or just communicating that it takes a while to process would go a long way for initial trust of the application, because we tend to think of food consumption as a discreet meal that happened at that one time, especially if people are used to the manual calorie-tracking application MyFitnessPal, which shows the full calorie count immediately.

The next design flaw in the GoBe application (Figure 2-13) is a little more complicated and took me a while to figure out. I was testing to see if the watch actually worked by meticulously tracking my calories manually and comparing the manual counts to the GoBe. Some big inconsistencies came up. I eat the same meal for breakfast every day, and it's exactly 400 calories. Some days the GoBe would report my calories at about 424, which is completely acceptable, but other days it would be off by more than 200 calories. The circumstances of my morning didn't change, but it would consistently be getting the counts wrong, and I couldn't figure out why, but I had a feeling that it might be in the time intervals that I selected. After a couple days, I realized that if I logged the meal as being consumed within a single 15-minute interval, the device would be surprisingly accurate; if I logged the meal as spanning two 15-minute intervals, it would log almost exactly 200 calories more than I ate. As it turns out, and this was later confirmed by the founder of the company, the time intervals are used not only for time logging, but also inform the calorie counting algorithm. In fact, it seems that the primary reason for the 15-minute based meal logging is to gauge the size of the meal, the logic being that if your meal takes longer, it's probably a bigger meal, thus higher calorie count. This design flaw gets even worse in higher-calorie meals, if I very quickly eat a 1,200-calorie burger and fries in a single 15-minute interval, and GoBe logs it as 800 calories, there's no way I'm going to ever trust the accuracy of the device, but if I say it took me a half-hour to eat, it's almost dead on.

Aside from the two quick fixes of holding off on the calorie reporting until full digestion and just simply asking if it was a big, medium, or smaller meal instead of indirectly asking how long the meal took, there's also the overall communication of the application that can get in the way. With wearable devices, and specifically health-related wearables, we have this assumption of absolute accuracy, but more often than not they're just accurate enough to give us an idea of the number of steps we've taken or the number of calories we've burned or consumed. In the case of the GoBe, the biological indicators it's working with do not mean the same thing for everyone. The device uses tissue impedance (similar to the Jawbone UP3) and, according to parent company HealBe, can take up to six months for its algorithms to fully learn how your body reacts to food. It simply can't work perfectly straight out of the box. This initial inaccuracy can present a problem for a $300 device. Is it better to sell the product to people and say, "Just hold on; it'll work better if you stick with it," or should they just hope people trust the thing and eventually everything will get better?

FIGURE 2-13
The GoBe application

Designing Fitness Trackers

Before we get in to the design of fitness trackers, let's talk about what fitness trackers actually do. Fitness trackers simply digitally encode some form of information that comes from our bodies, and then present that data to us in a meaningful way. More specifically, the fitness tracker records one or more biological indicators (bioindicators), translates that indicator into electrical signals, translates those electrical signals into data that can be stored, converts that data into some metric that is then recorded to some sort of database, and then displays these metrics in a way that possibly makes sense to you. There's a lot of processing that must happen to get that step count to you at the end of the day, and there's a lot of calculation and translation from between the bioindicator and data output that affect both the accuracy and communication of the information.

BIOINDICATORS TO SENSORS

When designing a new fitness tracker, after you determine what you want to measure, it's probably a good idea to begin with the *bioindicator*. Bioindicator is an ecological term that usually refers to a plant or animal species whose presence, absence, or behavior is an indicator of something bigger going on within its ecosystem. For example, naturalists might view the presence of bats as a bioindicator that the air and water in a particular area of an ecosystem is healthy. In the design of wearable devices, a bioindicator is any signal that we could use to monitor our activity or biological state. Examples of bioindicators for fitness trackers are things like swinging your arms in a certain pattern suggesting that you took a step, the changes in speed of blood moving through your veins suggest heart rate, or relative stillness over a certain period of time indicates that you're sleeping.

So, when you choose what you want to monitor—let's use steps, for example—you're going to want to put a sensor somewhere on the body that would best indicate that a step has been taken. The best place for the sensor would probably be in the sole of your shoe. Your foot has the most direct relationship to the activity that you're looking to monitor, but the placement isn't very practical because people wear different shoes all the time and don't want to have to remember to swap out their sensors. Where would be the worst possible place on the body to measure a bioindicator that would suggest taking steps? Probably your wrist. It's completely disconnected from the action, and if you're carrying something while you're walking, pushing a stroller or grocery cart, or doing anything else with your hand, it's going to be pretty inaccurate. A good middle ground is putting it on your hip or bra like the Fitbit One, but there are also problems with that such as losing it all the time, washing it in your clothes, and people not buying it to begin with because the wrist-based pedometers are now an industry standard.

SENSORS TO DATA

Given all that, it looks like we're putting the tracker on our wrist. What sensors do we need to at least attempt to get an accurate step count? We need a sensor that will measure movement on at least three axes, horizontal, vertical, and the z axis (depth), but because it's on your wrist, an appendage that moves around a lot, you're going to need to know which way is down so you know on which axes the movement you're recording is taking place. To measure movement along the axes, we'll use an *accelerometer.* An accelerometer measures acceleration in a given direction. For our tracker, we'll use a three-axis accelerometer to measure the acceleration of our wrist on the aforementioned axes. To understand which way is up, we need to use a *gyroscope.* A gyroscope is a sensor that calculates the orientation and rotation of our device. This is important because our arm changes orientation when we move it.

DATA TO METRICS

The easy part is getting the sensor data. Now that we have our raw movement data streaming from our accelerometer and corrected by our gyroscope, we need to convert that data into something that's useful to the person wearing the device. Figure 2-14 is my accelerometer data for walking around for about 10 seconds. What information can we assume from this data? Well for one we can see a pretty clear pattern of my steps. Let's just infer that each one of the higher x axis peaks (bottom line) to the following lower valley is a step because I'm swinging my arms in a repeated pattern, so those are steps (for this data set at least). Given my height, we can set a pretty universal standard for exactly where those peaks should land for walking; given my weight, we can assume how much energy I'm expending to walk and thus infer some sort of calorie expenditure based on that. If I'm running, I'd expect to see the peaks and valleys of this data to be much more different as well as faster. I could then assume from the data a higher rate of caloric burn.

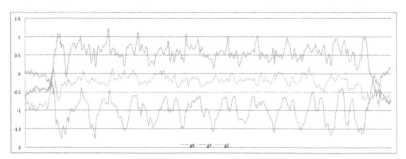

FIGURE 2-14
Graphed three-axis accelerometer output

METRICS TO INFORMATION

This is the fun part: giving structure to your metrics to turn them in to meaningful data. There are numerous options here, but we can begin with the basics: steps per day is a pretty standard way of measuring things. But I want to make this example a little more human, so instead of starting and stopping the day at midnight, let's start the day when you wake up and end it when you go to sleep. That way the number of steps per day matches how most people understand their day. On top of that we can do some basic math to provide more interesting information beyond the boring, "You hit 10,000 steps! Good job!" Let's take our sleep data and keep comparing it to the activity of the following day so that we can learn some trends. Our example tracker can then give reflexive information such as, "You got a lot less sleep last night; usually when you sleep that little, you're 30% less active" (Jawbone does this). Then we can be prescriptive by suggesting to our wearer to try to take an extra walk today to help make up for the steps she'll miss and try to get to bed a little earlier to catch up on the missed sleep.

Pulling It All Together

There are all kinds of ways to make fitness trackers interesting to users, and now it's pretty much a necessity if you want to differentiate your tracker from the existing huge field that all basically do the same thing. Your goal could be calorie burn, and that would probably be a lot more indicative of activity than steps alone, so why not have the day start and end when you wake up and go to sleep? Maybe sticking sensors in your ear turns out the be the absolute best place to put a pedometer and heart rate monitor. And what if the entire interface was aural? Throw GPS in there and give people feedback on how boring their day is, based on the fact that they did the exact same thing for the past four days. Trackers based on 10,000 steps have been around for a long time, but 10,000 steps doesn't mean the same thing for everyone. It's time to mix it up!

[3]

Smartwatches

We're still in the early days of smartwatches, which is odd because we've had them in some form or another since the late 1970s. With such a long history in the public consciousness, it's important to take a look back to see how we got to where we are today and what we can learn from the past. In this chapter, we look at the evolution of the smartwatch, beginning with the Pulsar Calculator Watch in 1975, all the way through contemporary smartwatches that work with our phones. We take a good, detailed look at what exactly these devices are adding to our lives (or taking away from them in some cases) so that we can move forward with design that is less influenced by existing patterns and trends, and more aligned with our current needs and values.

Later in the chapter, we get more specific about the cultural idea of smartwatches replacing full-sized counterparts and the inevitability of shrinking consumer technology, but for now, I want to hop in the time machine and take a look at some of the gems of our smartwatch past.

The First Calculator Watch

For most of their history, smartwatches have followed a similar pattern; they're made when the technology that drives their functionality has matured enough that it can be shrunk down to fit on someone's wrist. In most cases, the value of the watches came from the fact that they're replacing their full-sized counterpart. For example, in 1975, the Time Computing Company released the Pulsar Module 1 calculator watch (Figure 3-1), which could not only tell the time and date, but also add, subtract, multiply, divide, show percentages—it even had a little memory. The watch displayed six digits but could do calculations up to 12 digits with a floating decimal. With this kind of capability, it replaced "pocket calculators" that were about the size of four iPhone 6 Pluses stacked on top of one another, and often times needed to be plugged in to a power source. Even though you needed to carry around a stylus to work the diminutive keypad on the Module 1, its size and convenience were a big difference!

This watch followed the release of Pulsar's "Time Computer," which was the first digital watch. Both were enabled by advancements in light-emitting diode (LED) display technology, which was a beautiful glowing red and consumed a lot of power. This power consumption dictated a lot of design decisions on the device, chief among them was that the display remained off until the wearer pushed a big "Pulsar" button. The watch was also pretty bulky (15 mm thick) due to the size of the four batteries required to power it. It was advertised as having batteries that would last a year, but that was if you checked the time only 20 times each day and did only 20 calculations per day. The LED display was replaced a few years later by liquid-crystal displays (LCD), which are the black, always-on displays that you can still find on a lot of digital watches today.

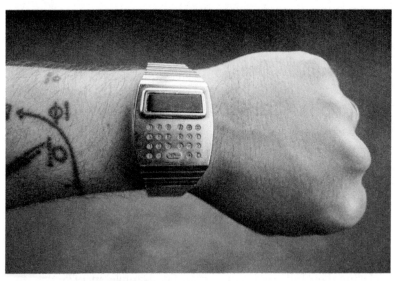

FIGURE 3-1
Pulsar Module 1

In 1975, which was a year away from Steve Wozniak's Apple 1 proto-type, the closest thing we had to computing power on a consumer basis were calculators. At the time, calculators themselves were cutting edge for the consumer market; before then it was slide rules and pencils. But by 1975, calculators were finally coming down in price enough so that they were affordable by pretty much everyone—you could get a basic calculator for $20 (about $90 in 2016 dollars). The calculators were huge compared to those you can buy today, but they weren't furniture. You could fit them in your pocket—if your pocket was big enough—and this is where the value of the Module 1 came through. The calcula-tor watches were originally sold at Tiffany for $3,950 ($17,394 in 2016) and were made out of 18k gold. They were so commercially successful that in early 1976 they came out with a stainless-steel version for $550 ($2,290 today), which seems rather insane now, but they were hugely popular. By having the watch on your wrist, along with the included pen/stylus combo, you were not carrying around a big "pocket" cal-culator, and the functionality was immediately accessible. Calculator watches came down in price in the 1980s when Japanese manufactur-ers got in the game, and they continued to be popular for decades.

The Microchip Explosion

By 1983, the microchip and integrated circuitry had gone mainstream and watch companies went nuts. In an article in the July 1983 issue of Popular Mechanics,[1] there was an explosion of watches that did more than tell time; you had video-game watches, FM radio watches, dictio-nary watches, digital chronographs, cheap calculator watches, a "jog-gers watch" that kept track of how far you ran, a watch that measured your pulse, a watch that could take your temperature, and a watch with which you could scroll through memos. Each watch was a technolog-ical marvel and work of art, but only one watch had stood the test of time as an icon: the Seiko TV Watch. Seiko's TV Watch is definitely not the most practical device in this chapter, but even today, it's just cool (Figure 3-2). It was worn by the iconic secret agent, James Bond, in the movie *Octopussy* and was the closest thing to the fictional Dick Tracey watch that anyone had ever seen.

---

1 Neil Shapiro, "Watches More Than A Good Time," *Popular Mechanics*, July 1983.

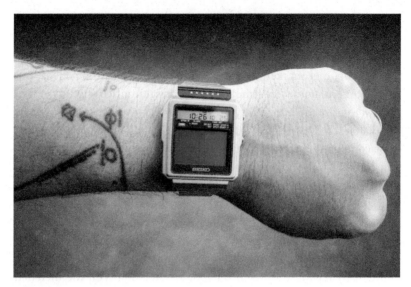

FIGURE 3-2
Seiko TV Watch

Aside from the obvious cool factor, the Seiko TV watch was particularly mind-blowing for the time because of its display. In 1982, almost all other televisions at that time were the bulky cathode-ray tube (CRT) televisions, whereas the TV Watch was a monochromatic "active matrix" guest-host LCD. Flat LCD or even the flat plasma-based televisions wouldn't be popular until the mid-1990s, with full market saturation not occurring for another 10 years after that. On top of being far ahead of its time in screen technology alone, it was the size of a watch! In 1982 the black-and-white nematic LCD displays that we associate with most contemporary digital watches were a cutting-edge technology. Just a few years before, we were looking at red LEDs.

A big question with the TV Watch was exactly how practical it was. To use the TV feature, you had to plug it in to the included television receiver, which was about the size of a Sony Walkman cassette player. On top of the extra hardware, it was $495 ($1,245 today), which was half the price of the stainless-steel Pulsars, but it still didn't sell very well (about 70,000 units in the first 18 months) because the receiver was awkward to use and the picture quality wasn't very good. As impractical and gimmicky as it might seem now, I don't think it was completely useless. We need to remember that at the time there wasn't the internet and there were three ways to get the news: newspapers, radio, and television. And the TV watch did deliver on two of those three.

The single-function (aside from normal watch functions) novelty watches continued into the 1990s. LCD video-game watches were cheap and popular among children, though as far as I can remember they weren't that great, and the games were very basic. Beyond the cheap game watches, there were some watches like Casio's JC-11, which was a pedometer and calculated how many calories you'd burn after you input your weight, age, and stride length. You also had to press a button to inform the watch whether you were walking or running, which is pretty cumbersome, and you would need to write the data somewhere if you wanted to keep it around for longer than that day. Another novelty watch that was somewhat popular in the '90s were the universal remote-control watches (Figure 3-3), which had an infrared LED on the front that could communicate with televisions and VCRs.

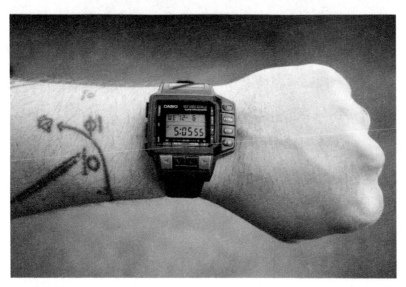

FIGURE 3-3
Casio remote-control watch

Early Wrist Computers

In 1983, Microsoft released the first version of Word, and Seiko released the first "wrist computer" with similar basic word-processing features called the Data-2000 (Figure 3-4). It was so named because it had exactly 2k of RAM, which allowed it to store exactly 2,000 characters' worth of memos. The watch came with a keyboard dock to manually punch in the information, and there was also a keyboard that you could purchase that had a printer attached. Two thousand characters might not seem like a huge deal, but 14 years before cell phones stored every number we've ever interacted with, we'd have to write them all down in pesky notebooks, and that was nowhere near as cool as the Data-2000.

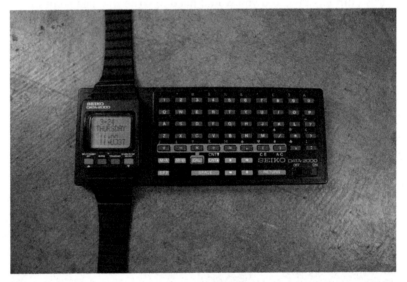

FIGURE 3-4
Seiko Data-2000 with keyboard

The memo function didn't quite make it a wrist computer; but the optional add-on processor that could snap onto the keyboard dock did. The keyboard with the printer added another 4k of RAM and was yet another attachment that worked with this larger keyboard that added 26k of ROM and ran Microsoft BASIC. The ROM "pack" was not trivial. The *New York Times* reported that it had "a calculating capacity equivalent to a personal computer" at the time and used the watch to display and save the data from your programs so that you could take them on the go. With all of its utility, this computer was not meant to be your only computer. 1983 was the same year that the Apple Lisa came out with the Lisa Office System, an early graphical user interface (GUI), and there's no way that the 1,400-pixel, dot-matrix computer watch running BASIC could compete on that level.

The second type of wrist computer was released in 1984 and was Seiko's RC-1000. The RC-1000 was similar to the Data-2000 in that it stored and displayed data, but instead of the terminal being a mini-keyboard dock, it was your computer via a serial cable. It worked with Apple, Commodore, Radio Shack, and IBM computers with included software and addressed a lot of the issues with data input that we still face with current smartwatches. *PC Magazine* described the Data-2000's keyboard as "a step from purgatory to hell," continuing to say, "The keys are too tiny to be called Chiclets and too unreliable to be called anything but profane."[2] In addition to getting around the tiny keyboard, it allowed you to keep your normal calendar on your computer and copy it with your watch via your computer, although it a far cry from just pressing a "sync" button like smartwatches today; it was through a bulky terminal interface and you had to copy and format each event, one by one.

Ecosystem Integration

The first "wrist computer" that could be considered user-friendly was probably the WristMac, Seiko's RC-4500 which came out in 1988 and could be considered the first "Apple Watch" (Figure 3-5). The watch itself wasn't that different from the RC-1000 in terms of functionality, but what set it apart was the interface software that shipped with it. You still had to manually copy your schedule one event at a time, but the interface was a step up from the command-line interfaces of the previous generation. For example, there was actually an image of the watch on the screen that you directly manipulated to set your alarms. The WristMac was so useful at the time that NASA gave it to the astronauts on Space Shuttle Atlantis for mission STS-37 in 1991, connecting the watches to a Macintosh Portable computer that was in communication with Mission Control via fax modem.

---

2 Winn L Rosch, "Wrist Computers: Wearing Your Screen on Your Sleeve," *PC Magazine*, June 25, 1985, 66.

FIGURE 3-5
The Seiko RC-4500 WristMac

The second smartwatch to go to space was the Datalink, a collaboration between Microsoft and Timex that was launched in 1994 (Figure 3-6). Various models of the Datalink were all over the International Space Station in the late 1990s and early 2000s, worn by both cosmonauts and astronauts. The astronauts got a lot of use out of the watches, as shown in an Expedition 1 crew log from 2001:[3]

> We have been working with the Timex software. Many thanks to the folks who got this up to us. It seems we each have a different version of the datalink watch, and of course, the software is different with each. Yuri and Sergei are able to load up a day's worth of alarms, but Shep has the Datalink 150, and this has a 5 alarm limit. So 2/3 of the crew are now happy. All this is a pretty good argument for training like you are going to fly—we should have caught this one ourselves in our training work on the ground.

3 Crew on the International Space Station (Alpha), "Expedition One January Crew Log," National Aeronautics and Space Administration, ship's log January 29, 2001 (*http://spaceflight.nasa.gov/station/crew/exp1/exp1shepjan.html*).

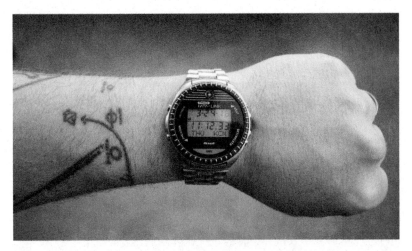

FIGURE 3-6
Timex/Microsoft Datalink

Aside from its astro-utility, the Datalink had a lot going for it. The early models used computer monitors to wirelessly flash data to the watch—you'd simply press the watch on to the screen of the monitor and it would flash black-and-white signals to a sensor that was on the front of the watch. The monitor method only worked with CRT monitors, though; when those fell out of favor, you would have had to use a serial cable with a flasher on the end of it. The Datalink is also particularly of interest because it was the first watch to automatically load events from an existing calendar software; it was compatible with Microsoft Schedule+, allowing it to work like current smartwatches that automatically sync your calendar.

A lot happened outside of smartwatches at this time in consumer technology. Pagers came in to play in the late '80s and began their exit in 2001; cell phones were beginning to become affordable, but they definitely weren't smartphones; and mobile computing was dominated by the Palm Pilot. Every task offloaded to the wrist computers was much better on a smartphone-sized touch screen, and you could take memos (pretty quickly with practice) right on the device. In 2003, we saw the familiar pattern of shrinking existing technology to wrist-size when Palm and Fossil combined forces to create the Wrist PDA (Figure 3-7). The Wrist PDA was pretty neat for the time; it was a fully functional

Palm Pilot with a mini-stylus in the buckle of the strap. As futuristic and useful as it was, it was also bulky, and Fossil stopped making it after two years.

FIGURE 3-7
Fossil Palm OS Wristwatch

Cell Phones and The Modern Era

The entire consumer technology landscape changed in 2007 with the launch of the first Apple iPhone. Before then, smartphones were clunky to use, mobile websites were an afterthought, and apps weren't really a thing. The smartphone as a window to the internet felt like a very small window, and you definitely weren't getting the full benefits of connected life. The first iPhone launched with the App Store, which brought fully featured mobile services and experiences to your phone that were explicitly designed for the device and proved to be extremely popular. The unfortunate side effect of this was that they were so popular that they pulled us away from other aspects of our lives, and because they were in our pockets, they had access to our psyche on a level that has not previously existed. This is the ecosystem that gave birth to the modern era of smartwatches.

Although there were definitely other notification-based watches being produced and developed at the time, the Pebble Smartwatch really set the pace for the group when it started shipping in 2013 (Figure 3-8). Initially funded by what was the most successful Kickstarter campaign at the time (more than $10 million), the watch was incredibly popular and sold a million units in its first two years. Pebble's relatively low price and the fact that it worked with both Android and iPhones was a major factor in its popularity, but its biggest contribution was that it received a lot of attention and thus moved smartwatches away from the niche market and into mainstream consumer electronics.

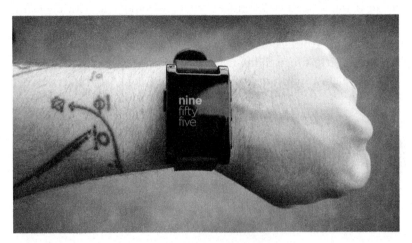

FIGURE 3-8
First-generation Pebble Smartwatch

Not too long after the Pebble, the smartwatch went mainstream, with Google and Apple getting into the picture, and their approaches to their watch operating systems could not have been more different. There were certain things that both competing operating systems did, such as automatically sync with your calendar, track your steps like any fitness wearable, and of course display any notification that might come across the phone's lock screen; but that's pretty much where the similarities ended. The companies diverged greatly in the details of their platforms, and each utilized the strengths of its mobile operating systems, iOS or Android, to shape the watch experience.

Google launched its Android Wear platform with a couple of watches in 2014 with a strong emphasis on the existing Android core ecosystem as opposed to third-party applications. Android watches still displayed all of the notifications that would otherwise pop up on your phone's lock screen, but the primary strength of the operating system was definitely the integration with Google's personal assistant tools. Google Now is Google's intelligent personal assistant that uses contextual information such as geolocation and time, combining it with Google's massive amount of information that it gets from your Gmail account, searches, calendar, lists, and other sources to deliver a seemingly serendipitous experience. For example, when you woke up in the morning, as you were getting ready, it would keep you updated with the amount of time it should take you to get to work based on local traffic; or if you walked into a grocery store, it would automatically bring up your grocery list. Another strong feature of Android Wear was its "Okay Google" voice-controlled virtual assistant technology (equivalent to Apple's Siri), which didn't need to be physically activated by touching the watch. You just needed to say, "Okay Google" and start talking.

FIGURE 3-9
Moto 360 Androidwear Smartwatch

Apple took a very different path with its watchOS for the Apple Watch, which definitely played to Apple's strengths by utilizing third-party developers to create much of the functionality. The watchOS user interface (UI) is centered around what Apple calls *glances*, which are basically previews for the individual watch apps, they're accessed by swiping up on the home watch face screen and are meant to give you quick access to app functionality that you use more often (you choose which apps appear in glances). The odd thing about Apple's launch of its watchOS in 2015 was that the third-party applications had almost no access to the watch hardware such as the accelerometer, heart rate sensor, Taptic Engine (haptic), microphone, digital crown and other interface elements such as watch complications (more on those later). It took Apple five months to release a second version of watchOS that gave access to these features, which likely affected the initial reception of the new device and suggested a rushed launch.

Frameworks for Smartwatch Interaction

Now that you've seen how smartwatches have evolved into what we have today, let's take a look at our entire digital device ecosystem to understand the relative positioning of smartwatches alongside all of the other devices in our lives. I like to think of these devices on a spectrum of intimacy based on the prerequisites for, and context of use. For example, on the less intimate end of the spectrum are desktop computers. To use your desktop computer, you need to be in the same room, turn it on, and open an application to engage with it. Laptops are similar except you don't need to be in the same room. Toward the more intimate side of the spectrum, you have a smartphone, which you must take out of your pocket, unlock, and open an application. The primary differentiator is that your phone is with you everywhere you go and notifies you of incoming messages. To engage with smartwatches, however, you simply look down at your wrist, making their access to or psyche nearly direct, which has the potential to disrupt people's lives in a negative way.

A useful way to look at the different types of engagement with our devices across the intimacy spectrum is to look at our interactions as *intentional* versus *reactive*. For my desktop computer, the interactions are intentional because I am actively engaging with the device. They also are completely intentional due to the prerequisites of my taking steps to physically engage with the device. Modern smartwatches do not have any physical barriers that prevent you from using them; when your wrist vibrates and/or beeps, it's almost impossible not to look down at it.

The problem with how smartwatches are used right now is that the interactions are almost entirely reactive, and the result is that we're taking control away from ourselves (as the watch wearer) and handing it over to whomever might be notifying us. The responsibility to protect the consciousness of the consumer then falls to the designers and developers of the devices and the applications that live on the watch. I can ignore a vibration in my pocket when I choose to do something more meaningful than to check my phone; I cannot ignore a vibration on my wrist and a flash of light in my field of vision. In the next couple of sections, I discuss how we can mediate this in the design of applications for smartwatches as well as the design of the devices themselves.

Designing for Smartwatches

In November 2014, I was a design mentor for the first official Apple WatchKit hackathon. It was a great opportunity to get a handle on the exact capabilities and limitations of the new watch, to get an idea of what people wanted to build, and to think about how to best go about doing that, given that the Apple Watch was going to inevitably change the way we thought about smartwatches. My biggest piece of advice to the designers and developers at the hackathon was that watch applications don't need to be more simple than phone applications because the screen is smaller; they need to be more simple because of the intimate nature of the device. Because of the physical position of the watch on someone's body, it's inevitably going to have more access to the user's attention, so the applications need to be paired down to ensure that they're less of a disruption. Smartwatches aren't mini-phones that you wear on your wrist, and you need to look at the specific advantages of that form factor. If the watch is only serving as a reminder to look at our phones, we have failed. Sure, there's a slight advantage to not having to take your phone out of your pocket to see who just responded to your Facebook post, but it's not worth the 500 times that your wrist vibrates each day.

Looking at where the watch is on our body and what the watch is physically capable of doing, there are two types of applications that can be truly useful, aside from the omnipresent notifications: *active-focused applications* and *passive-focused applications*. Active applications are initiated by the needs of the user and would be something like an application to trigger a Bluetooth-enabled camera shutter or maybe take a voice memo, something that the user is actively choosing to do. These are the applications for which the watch is primarily used for input, and the size of the device can really be an impediment. Smartwatch faces are roughly one inch square in size, and because of this, they're not all that great at targeted touch input like we have on smartphones. If you're going to put buttons in an area that small, you're really only going to have two of them along with any contextual information. Beyond buttons, short swiping motions work well because they don't require a lot of precision. Both Google and Apple operating systems heavily rely on swiping for most of their navigation.

Capture, GoPro's Apple Watch app (Figure 3-10), is a great example of properly designed touch-based application. The app has two screens, one is a giant red button to start and stop recording. If you swipe to the right, you can see a low-resolution preview of what's in the camera's frame. GoPro's app isn't just great because of its paired-down interface, though. It's a great application because it's contextually appropriate: a small screen strapped to your wrist is the perfect place to have controls for a camera that's designed to be used in physically active situations. Beyond the touch screen, the watch has a limited set of sensors that could prove to be pretty useful. For example, the accelerometer/gyroscope in Apple Watches could be used to control the roll, pitch, and yaw of a drone by mimicking the movement of your arm, creating a one-to-one ratio to movement of the drone. Evernote does a good job of sensor-based input on its Apple Watch app by offering text input using the microphone by converting speech to text to create short notes through the watch.

FIGURE 3-10
GoPro's Capture app for Apple Watch

I believe that the power of the smartwatch does not lie only in timely notifications or contextually appropriate active applications, but in a persistent peripheral reminder of items that progress throughout the day. If we look at the more traditional (precomputer) role of the watch, it serves as a narrowly focused dashboard that we use for consistent awareness of time. This peripheral awareness aids in contextual decision-making. Therefore, the opportunity of the smartwatch is to provide a custom dashboard of relevant information that is now available to us. Using traditional horological terms, anything on the watch face that is not the time is called a *complication*. On traditional watches, complications are usually things like a little window that displays the date, or the phase of the moon; or on chronograph watches, a stopwatch. On smartwatches, complications are not constrained to time-based information; they can be a dashboard for any relevant contextual information we have on our phones, such as weather, physical activity, financial data, sports scores, and our schedule.

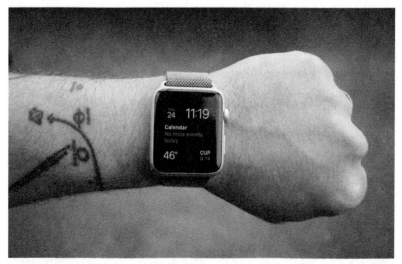

FIGURE 3-11
Apple Watch with complications

An example of these passive complication-based applications using financial data would be an app that would pull my spending data from my bank and my credit cards and compare it to my budget. A glance at my watch could tell me that I can definitely go to happy hour after work because I'm $40 below my weekly discretionary spending budget. If presented passively, this information can help me make decisions using information that I don't even need to consciously acknowledge but can be in my periphery and I can be aware of this information without being distracted by it. Luckily, when Apple released watchOS2, it added the capability to create third-party complications.

Designing Smartwatches Themselves

We're still definitely in the early days of smartwatch design. Pebble took the lead in defining the functionality and expectations of the current generation of smartwatches, but now that the technology has gone mainstream, I'm expecting a bit more diversity than just the Apple and Android platforms. The first Apple Watch prototype was an iPhone strapped to the wrist of the researchers, and this decision alone comes with a lot of assumptions around the design of the rest of the watch. From that point forward, they knew that the watch would more or less take the form of a tiny iPhone that you wear on your wrist. The next generation of smartwatches will likely challenge the assumption of the smartwatch extending technology that was developed for the phone.

Screens on your wrist are costly. They're expensive to make and they consume a lot of power. If we take a step back and look at how people currently are using their smartwatches (fitness tracker and notification center) and keep in mind what smartwatches are good at, we can design a watch from the ground up that can play to the strengths of the form factor and body position. I'm developing a watch with a colleague of mine that's a mechanical smartwatch and looks similar to a classic chronograph, but the physical dials of the complications are tied to your digital life instead of time-based information. Aside from staying out of your way, mechanical smartwatches don't need to turn off when you're not glancing at the screen, because they consume significantly less power than a screen display.

The closest thing to this that's available are the Withings Activité watches (Figure 3-12) that are fitness trackers that look like a mechanical watch with a single subdial to display your progress toward an activity goal. Though you're not getting notifications or a ton of information on the watch, you're wearing something on your wrist that looks like a classic watch and has a battery that lasts about eight months. In addition to the Activité, there are a handful of other experiments that are being tested in the market. For example, Fossil is selling a handful of what it calls nondisplay smartwatches, which are mechanical watches with a Bluetooth transmitter strapped to the back that flashes a colored LED when you get a certain notification. There are also "smart straps" that you can connect to a traditional watch that adds similar notification and sensor functionality.

FIGURE 3-12
Withings Activité Steel

Pulling This All Together

The most important thing to keep in mind when designing software for smartwatches or smartwatches themselves is that there are certain things that a smartwatch is more appropriate for than a smartphone, and those things are usually passive. There are a ton of smartwatch applications that just exist because they can, but they aren't really adding anything to their service by existing on the wrist. If you're designing one of the few active applications that can truly take advantage of the sensors and body positioning of the watch, you have a limited set of options for input. The vast majority of genuinely useful applications for current smartwatches will take the form of a small complication on the watch face and be little more than an item on a dashboard of your digital life.

[4]

The Glass Experiment

A Truly Wearable Computer

Google Glass, and what happened with Google Glass, has to be one of the most interesting stories in the history of wearable devices. The device came from a surprisingly long history in terms of functionality. It became a major cultural touchpoint and ultimately taught us a lot about where we are in terms of public acceptance of certain technologies. This entire book is about *wearable computers*, which are basically any devices strapped to your body that perform computation, but Glass belongs to a group of wearables that are truly wearable computers, a general-purpose computational platform that is designed to be worn on the body.

Glass is particularly interesting when you think about its relationship to computers as a whole. As I mentioned in the beginning of the book, we're approaching the point at which technologies are small and powerful enough to no longer be restricted to what we now consider their traditional forms. Without significant constraints in terms of the size of the technology, we're free to create computers that are designed from the ground up with our bodies in mind, and a truly wearable computer might make the most sense as the ideal form factor of computers as a whole.

EARLY WEARABLE COMPUTERS

Before computers had an established form factor of terminals with typewriter-style keyboards and big flat monitors, there were a handful of people who built devices that we wouldn't recognize as computers today. The first of which is Morton Heilig's Sensorama Simulator (Figure 4-1), which he patented in 1962 and was sort of an early virtual reality system designed for immersive learning.[1] The Sensorama Simulator was one of the first devices specifically designed around our bodies and our senses, including sight, hearing, touch, and even smell. A couple of years later, Ivan Sutherland created an early head-mounted, three-dimensional display nicknamed The Sword of Damocles,[2] which was specifically designed to take advantage of head movement relative to displayed information on the display, creating the illusion of dimensionality of a virtual object.

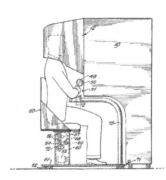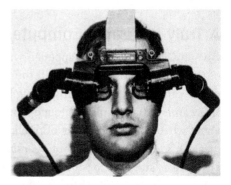

FIGURE 4-1

The Sensorama Simulator and the Sword of Damocles

The next big push in the direction of a fully wearable computer came from the MIT Wearable Computing Project in the mid-1990s. The group, colloquially known as *The Borgs*, focused on developing wearable computers that were meant to be worn constantly.[3] It was here that Thad Starner, who would later become the technical lead for Google Glass, developed the earliest truly wearable computers. In 1993, Starner and fellow Borg Brad Rhodes coded the *Remembrance Agent* (Figure 4-2), which was a note taking and retrieval program that pulled

---

1 Morton L Heilig, Sensorama Simulator, US Patent US 3050870 A, filed January 10, 1961, and issued August 28, 1962.

2 Ivan E Sutherland, "A Head-mounted Three Dimensional Display," *Proceedings of the December 9–11, 1968, Fall Joint Computer Conference, Part I on - AFIPS '68 (Fall, Part I)*, December 1968, 757–764. doi:10.1145/1476589.1476686.

3 Steve Mann, "Smart Clothes: The MIT Wearable Computing Web Page," last modified December 18, 1995. (*http://www.wearcam.org/computing.html*).

information from a database of notes from previous conversations.[4] The first reliable wearable computer that Starner built was Lizzy 1, which was a backpack-based system with a Twiddler one-handed keyboard (see Chapter 1) and an early head-mounted display where Remembrance Agent queries would be displayed. Beginning with the Lizzy 1, Starner wore his computers at all times for roughly 20 years, only taking them off to sleep, shower, and when he married (Figure 4-3 shows the Lizzy 2).[5]

```
Buffers Files Tools Edit Search Help
email and starts editing a file, the RA automatically changes it
recommendations accordingly.  These suggestions are presented in the form
of one-line summaries at the bottom of the screen.  Here they can be easily
ignored, or the full text of the suggestion can be brought up with a single
keystroke.

Most applications for augmenting human memory, e.g. those developed by
(Jones 1986) and (Lamming & Flynn 1994), have concentrated on
-----Emacs: remembrance-agent.txt   11:44am 0.05    (Text Fill)--L27-- 9%---
1      0.41 Felice Napolitan 24 Jan 96 | Remembrance Agent talk/discussion
2      0.33 Brad Rhodes .... 25 Jan 96 | Remembrance Agent available for Be
3      0.31 Sumit Basu ..... 14 Dec 95 | Re: keystrokes...
4      0.16 fellowship | testarne | Oct 23 1995 .....orientation forms in o
 *remem-display*
```

FIGURE 4-2
The Remembrance Agent interface

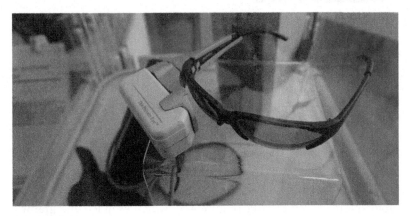

FIGURE 4-3
Thad Starner's Lizzy 2

4 Bradley J Rhodes and Thad Starner, "Remembrance Agent: A continuously running automated information retrieval system," *The Proceedings of The First International Conference on The Practical Application Of Intelligent Agents and Multi Agent Technology*, 1996, 487–495 (*http://alumni.media.mit.edu/~rhodes/Papers/remembrance.html*).

5 Alex Spiegel, and Lulu Miller, "Computer Or Human? + Thad," *Invisibilia*, NPR, February 12, 2015 (*http://www.npr.org/2015/02/13/385793862/computer-or-human-thad*).

The Development of Google Glass

The Google Glass project began in 2010, and according to Google's chief evangelist, Gopi Kallayil, the first Google Glass prototype took about 90 minutes to construct. He described the process at the Dreamforce conference in 2013 (see also Figure 4-4):[6]

> So, it looked like this: a regular laptop computer, like the one you'd carry in your hands, put in a Google backpack that you can buy from the Google store, with wires protruding out of it, mounted on ski goggles, then you can buy off-the-shelf components like still cameras, video cameras, and sticky tape that you get from the supplies cabinet.

FIGURE 4-4
Thad Starner wearing an early Glass Prototype

After the initial proof of concept, they swapped out the laptop for an Android phone that they taped to the side of a pair of safety glasses. This was the *Pack* prototype, and it was completed in December 2010 and weighed 3,330 grams.

After the Pack prototype, the group focused on making the prototypes lighter and more compact so that they could be worn completely on the head. The first prototype that didn't require a backpack was the *Ant* prototype in March of 2011 (Figure 4-5), which weighed 167 grams.[7] Next was the *Bat* prototype (not pictured), which got rid of the phone case and weighed 115 grams. The *Cat* prototype, built in May 2011, is

---

6 Mark Billinghurst and Hayes Raffle, "The Glass Class: Designing Wearable Interfaces," course taught at the CHI 2014 conference, Toronto, May 1, 2014.

7 Clint Zeagler, Thad Starner, Tavenner Hall, and Maria Wong Sala, *On You: A Story of Wearable Computing*, Exhibit at the Computer History Museum, Mountain View, CA, June 30–Sept. 20, 2015.

the first prototype to use the optics design that would later be in the production version of the Glass. This version weighed 153 grams. The Ant, Bat, and Cat prototypes all were built by repurposing Nexus One phones.[8]

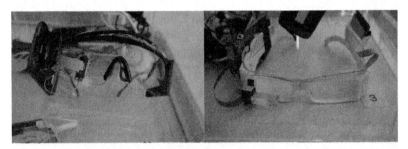

FIGURE 4-5
Ant Prototype (left) and Cat prototype (right)

The *Lennon* prototype (Figure 4-6), completed in April 2011, was a minimalist version of the Glass and was built in parallel with the Ant, Cat, and Dog prototypes. Lennon was built to determine how light the device could be and still be useful. Two *Dog* prototypes, *Dog Metal* and *Dog Plastic*, were built in June 2011 and were the first prototypes in the animal series of prototypes that used higher-end computer boards instead of repurposed phones.

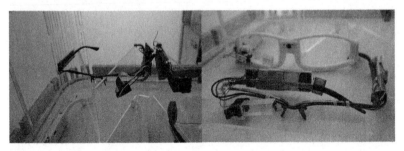

FIGURE 4-6
Lennon Prototype (left) and Dog metal and plastic prototypes (right)

---

8 Clint Zeagler, Thad Starner, Tavenner Hall, and Maria Wong Sala, *Meeting The Challenge: The Path Towards a Consumer Wearable Computer* (Georgia Institute of Technology 2015).

The *Emu* prototype (Figure 4-7), weighing in at 68 grams, was built in September 2011 and was the first prototype that featured an enclosure. It also was the first prototype that used bone conduction for audio. The *Fly* prototype was built in October of 2011, less than 11 months after the project began. It was the first fully functional prototype that resembled the more rounded final design.

FIGURE 4-7
Emu prototype (left) and Fly prototype (right)

With the *Gnu* (December 2011), *Hog* (February 2012), *Ibex* (May 2012), and *Koala* (October 2012) prototypes, the group refined the designs and began user testing. The Hog was the first prototype that was able to be worn publicly, since the project was announced in April.

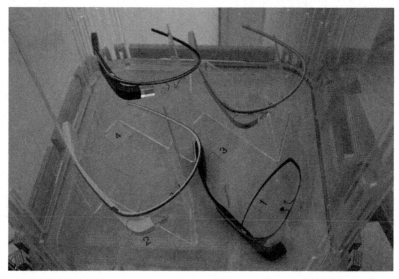

FIGURE 4-8
The Gnu (1), Hog (2), Ibex (3), and Koala (4) prototypes

One Day

On April 4, 2012, Google uploaded a video to YouTube called "Project Glass: One Day..."[9] The video didn't show what the device looked like; rather, it was a concept video that showed what a day might be like with augmented reality (see Figure 4-9). Some of the use cases that were demonstrated were text messaging via voice with the protagonist keeping both hands on his sandwich, a geolocated notification that there was a train delay, and a prompt to take a different route (with turn-by-turn directions). The man even goes into a book store and asks Glass where the music section is located, and it shows him! The end of the video shows the protagonist serenading a young woman with a ukulele via video chat while he shows her his view of a sunset.

FIGURE 4-9
Still from the "One Day..." video

The video is accompanied by a post by the "Project Glass" account on Google+ stating, "We think technology should work for you—to be there when you need it and get out of your way when you don't. A group of us from Google[x] started Project Glass to build this kind

---

9 "Project Glass: One Day..." Google, Published on YouTube April 04, 2012 (*https://www. youtube.com/watch?v=9c6W4CCU9M4*).

of technology, one that helps you explore and share your world, putting you back in the moment."[10] This is when the internet pretty much exploded. It was all over every blog and news site; it was all anyone talked about.

At the 2012 Google I/O conference, Google cofounder Sergey Brin demonstrated Glass for the first time by having a team of skydivers jump out of a plane above the venue wearing Glass and broadcasting the video feeds live (Figure 4-10). He then announced the first sale of the devices: the US-based developers who were present at the conference could preorder the device for $1,500, and they'd get the devices early 2013. For the rest of 2012, Glass was front and center in the tech news cycle. It was named by *Time Magazine* as one of the best inventions of the year and even popped up on the runway during New York Fashion Week. Glass was everywhere.

FIGURE 4-10
Sergey Brin talking to skydivers demonstrating Glass

The first Glass Explorer editions shipped in April 2013, almost exactly a year after the initial announcement, and were finally out in the world. They didn't have a great deal of functionality yet, but they could do certain things like record video, take pictures, get notifications, and

---

10 "We think technology should work for you—to be there when you need it and get out of your way when you don't," *Google Glass Blog*, April 4, 2012 (*https://plus.google. com/+GoogleGlass/posts/aKymsANgWBD*).

display turn-by-turn directions. People were using Glass in new and interesting ways: surgeons were using it to record and broadcast surgeries; citizen journalists livestreamed protests in Istanbul; and athletes trained with them. Even though the devices were interesting and potentially useful, they were largely overshadowed by public criticism around privacy concerns and tech elitism.

PUBLIC BACKLASH

The public backlash against Glass began before the devices were even in the public's hands. In March, 2013, a month before the product was released, West Virginia signed a bill to ban them while driving,[11] and the first public place to ban their use, the 5 Point Cafe in Seattle, put up signs warning people not to wear the device.[12] This was followed by many other establishments. March 2013 is also when the group Stop The Cyborgs formed to combat Glass and other technologies, distributing signs and stickers stating, "Google Glass is banned on these premises" (Figure 4-11). Most of the concerns were, of course, around privacy—people don't want to be filmed by other people in public. Another early point of concern was the potential use of facial recognition applications. This eventually prompted Google to publish an update saying that no facial recognition applications would be approved without having strong privacy protections.[13]

The backlash against Glass came to a head in February, 2014 at Molitov's, a dive bar in San Francisco's Haight-Asbury District. Tech writer Sarah Slocum was wearing her Glass like normal. When she first arrived at the bar, people were asking her about it and everything seemed fine; just normal people being curious. A few minutes later, though, things began to become more hostile. First, it was a woman turning to her, flipping her off, and yelling "F Google," this was followed shortly afterward by that same woman yelling, "Get out of here,

---

11 West Virginia. Legislature. House of Representatives. Committee on Roads and Transportation then the Judiciary. A BILL to amend and reenact §17C-14-15 of the Code of West Virginia, 1931, as amended, relating to traffic safety; specifically, establishing the offense of operating a motor vehicle using a wearable computer with a head-mounted display. H.B. 3057. 2013 Reg. Sess. (March 22, 2013).

12 "Google Glasses BANNED," *The 5-Point Café*, March 11, 2013 (*http://the5pointcafe.com/ google-glasses-banned*).

13 "Glass and Facial Recognition," *Google Glass Blog*, May 31, 2013 (*https://plus.google.com/ u/0/+GoogleGlass/posts/fAe5vo4ZEcE*).

b!@ch." That's when things turned violent. People started trying to rip the glasses off Slocum's face and throwing bar rags at her. One woman told her, "You're killing the city," and finally pulled the device off her face. Slocum eventually got it back, but when she chased the person out of the bar, someone else stole her purse and her cell phone.[14] Over the next few months, there was an attack in the Mission District in San Francisco which involved a woman with a mohawk running up to a writer wearing the device, screaming "Glass," and ripping them off the man's face before smashing them on the ground.[15] One other report describes someone in LA's Venice Beach being robbed of his Glass by someone using a taser.[16]

FIGURE 4-11
Anti-Google Glass sticker from stopthecyborgs.org

14 Sarah Slocum, "Google Glass Assault and Robbery at Molotov's Bar, Haight St. February 22, 2014 (warning: profanity)," *I Love Social Media:*, March 21, 2014 (*http://ilovesocialmediainc.blogspot.com/2014/03/google-glass-assault-and-robbery-at.html*).

15 Adario Strange, "Another Google Glass Wearer Attacked in San Francisco," *Mashable*, April 13, 2014 (*http://mashable.com/2014/04/13/google-glass-wearer-attacked/#IUBoJNXakPqX*).

16 Kia Makarechi, "Google Glass Wearer Robbed at Taser Point," *Vanity Fair*, April 16, 2014 (*http://www.vanityfair.com/news/tech/2014/04/google-glass-wearer-robbed-at-taser-point*).

So how did Glass become such a hated device? You can certainly put some of the blame on privacy concerns; that definitely had a lot to do with it. There's also "The Walkman Effect," a concept inspired by the Sony Walkman in the 1980s, by which people can be isolated in public spaces by technology. German psychologist Rainer Schönhammer observed:[17]

> This seems to interrupt a form of contact between "normal" people in a shared situation, even if there is no explicit communication at all. People with earphones seem to violate an unwritten law of interpersonal reciprocity: the certainty of common sensual presence in shared situations.

Though social isolation has been a negative talking point since newspapers were invented, Glass was a lot more high profile than most devices and had a lot of social hurdles to get over.

GLASSHOLES

Aside from privacy concerns and the groups that freak out every time something new happens, the third part of the controversy are the A holes that bought the device, or, as they became known, the Glassholes. The overt hatred of Glass users was caused by a perfect storm of technology hype, exclusivity, and income inequality. Google decided to roll out the Explorer program very slowly, initially to a small group at the I/O conference, and then through a contest, personal invites from the original explorers, and eventually to the general public. But even then, you had to have an extra $1,500 lying around to get one. The problem was that Glass was all over the news as a crazy new technology, everyone wanted to try it, but very few people could.

Unfortunately, most of the people who *did* have the credentials and means to get their hands on it were young, wealthy, white male tech workers in San Francisco and New York, a demographic that is not particularly known for their social skills or tact. Early Glass explorers caused a lot of problems by refusing to take off the device when asked, inappropriately filming people, and being rude to people asking them about it. It got so bad that in February 2014, Google published a list of

---

17 Rainer Schönhammer, "The Walkman and the Primary World of the Senses," *Phenomenology + Pedagogy* 7 (1989): 127–144.

dos and don'ts for using Glass, directly addressing user behavior and instructing people to not "be creepy or rude (aka a 'Glasshole')" and to "Respect others' privacy, and if they have questions about Glass, don't get snappy."[18]

Nobody wants you filming them all the time, and I get that, but there were bigger issues at play than etiquette. Though the violence in San Francisco involved people using Glass, the attacks weren't about the device itself, they were about the growing frustration with the tech industry and gentrification (see Figure 4-12), a problem that existed well before and after the Glass. The attackers didn't yell, "Stop recording me." They yelled, "F Google." The Mission attack was right after an anti-Google protest in the neighborhood; the device was merely a symbol. Even though the attacks were more about San Francisco's wealth inequality problems than the device, the attacks were even more negative press for the device.

FIGURE 4-12
Anti-tech-workers posters in the Mission District, San Francisco

18 "Glass Explorers," *Glass Press*, February 15, 2014 (*https://sites.google.com/site/glasscomms/glass-explorers*).

NOT A FAILURE

On January 15, 2015, Google announced the end of the Explorer program and stopped making the devices available to the public. So, was Glass a failure? Yes and no. The *device* was absolutely not a failure, it's a great piece of technology that is very useful (as we'll see in Chapter 5) and had a lot of very smart people behind it. What we have to remember here is that Glass was never a commercially released product; it was a beta version, plain and simple. Where Glass *did* fail was how the company handled the launch and publicity of the Explorer program.

One of the only people from Google[x] to talk publicly about this failure was Astro Teller at the closing keynote of South by Southwest Interactive in 2015, two months after Google shut everything down. "We allowed and sometimes even encouraged too much attention for the program," Teller admitted "Instead of people seeing the Explorer devices as learning devices, Glass began to be talked about as if it were a fully baked consumer product." He went on to say "we learned a lot from the very loud public conversations about Glass and will put those learnings to use in the future. I can say that having experimented out in the open was painful at points, but it was still the right thing to do."[19]

Pulling This All Together

Google's Glass Experiment was a well-meaning attempt to build a truly wearable, multifunctional computer platform, and the company created an incredible device that was derailed by the greater social context of the technology industry and poor management of the public image of the device. This type of platform has been a dream of the wearable device community for a very long time and will definitely not go away. Though we might not be wearing Google Glass devices any time in the near future, there are certainly technologies that are coming online that will expand upon the general concept and will certainly achieve the goal of a truly wearable computer. We discuss these in Chapter 5.

---

19 Astro Teller, "Moonshots and Reality," Speech, South By Southwest, Austin, March 17, 2015 (*https://youtu.be/3e0c0rL00bg*).

[5]

Wearable Cameras

Cameras present a huge opportunity for the wearables industry, yet they're held back by some of the most difficult social design constraints. What makes cameras highly valuable as an input device is the same thing that makes them socially dangerous: the amount of data that they collect. You have a sensor that collects enormous amounts of highly detailed visual information, but the collection of such detailed information is far from socially acceptable because of privacy concerns. The second issue that comes from this massive amount of data is what to do with it—how can we make this information useful and easily accessible when there's so much of it. In this chapter, we discuss why and how people use wearable cameras today, design decisions that mediate social issues around wearing a camera in public, what to do with the massive amounts of data that needs to be stored and used, and opportunities for future services that we can build from this type of information.

We typically use cameras in three different ways: reactively, actively, and passively. Reactive capture is when we see something we want to take a picture of (typically pull out our phone) and quickly try to save the image; this is how most people use cameras. Active capture is when you turn the camera on proactively to capture something; this differs from reactive capture because you're not reacting to the moment or subject of capture, and this is the way we typically use video cameras like the GoPro. The third way, passively, is when the camera is always running and it captures whatever is happening regardless of the intention or moment; security cameras are typical of this use category, and in the wearable world, this is where life-logging cameras like the Narrative Clip fits in.

Narrative: Design for the Photographed

Martin Källström founded Narrative in 2012 to create a camera that captured the smaller moments in life. Until the first Narrative Clip (Figure 5-1), nearly all photographs were captured reactively, meaning that we photographed moments that we felt compelled to preserve—moments that seemed significant enough to pull out a camera and photograph. Martin wanted a camera that would capture life between those moments, a camera that would document and preserve the parts of our lives that didn't seem significant at the time, but end up being incredibly meaningful in that they're more representative of what life is really like. With this goal in mind, Martin set out to design a wearable camera to constantly take photographs but also melt into the background.

Passive photography comes with some very significant social issues, mainly that you're wearing a camera that's photographing people all the time, and that freaks some people out. Martin knew this going into the project and decided to design the camera for the subject of the photograph instead of the photographer wearing the camera. The design emphasized two major principles: subtlety, to design a camera that didn't pull attention away from the moments; and honesty—this isn't a spy camera that you use on your friends. If people are being photographed, they should know it.

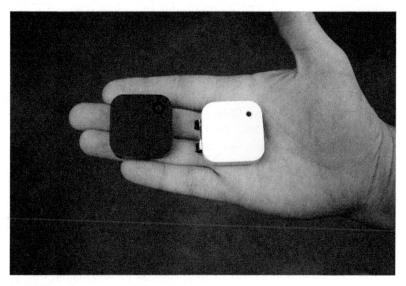

FIGURE 5-1
Narrative Clip version 1 (right) and version 2 (left)

The need for subtlety and honesty played out in the design process in very interesting ways. The first prototypes of the Narrative Clip were round with the lens in the center, so the device looked more like a camera lens. This would be a signal that it's a camera and speak to the honesty part of the design. However, when testing the prototype, designers found that the concentric circles too closely matched the pattern of the human eye, and when it was out in the wild, people continually felt the need to make eye contact with the device, making it not so subtle.

The next iteration of the camera was square (and eventually the company moved the lens to one of the corners), which made it less distracting, but it also looked less like a camera, swinging it away from the honesty goal. To make it look more like a camera, designers decided to take design cues from the most prevalent camera in the world at the time, the one on the back of the iPhone 4, and thus placed a silver ring around the lens on the case. Narrative updated this reference for the Narrative Clip 2 to reference the iPhone 6 with a larger lens opening which is more flush with the case, getting rid of the shiny edge of the original version.

My favorite design decision for the camera is the power button, or, more accurately, the absence of the power button. To really drive home the honesty of the design, the designers decided to forego an off button so that you could never reassure anyone that the camera you're wearing is actually off. In fact, the power mechanism is the accelerometer in the camera body that turns the camera off when the camera is still and face down on a table, enforcing the honest behavior through design. All of these design decisions shaped a device that's beautiful for its aesthetics as well as its commitment to the context of its use.

GoPros for Everyone!

Given their small size, durability, and robust offering of mounting accessories, GoPros have the ability to go everywhere people do without getting in the way. In addition to action sports, they're used by the Discovery Channel to film television shows, they're used by surgeons to document surgery, and they're heavily used by soldiers in combat. Because they're active capture cameras, they don't need to address the social problems that other wearable cameras have. The Google Glass received a lot of criticism (see Chapter 4) by proposing a use pattern of always wearing the device, causing a public outcry that ultimately led to the product not being offered commercially. Narrative didn't have these problems because it was designed from the ground up for the people being photographed. GoPro seems to have completely side-stepped this by establishing that it's not meant to be a passive, always-on camera, but used only for discreet activities.

Nick Woodman began experimenting with wearable cameras in 2002, when he built a wrist strap to hold a disposable camera so he could more easily capture his surfing adventures. This later evolved into GoPro's first product, the original HERO (Figure 5-2) which was a 35mm film camera inside of his custom-designed waterproof housing with a wrist strap. GoPro's first video camera was released in 2006 and could only record 10-second video clips. GoPro slowly evolved its cameras in to the HD HERO, which launched in 2010 and was a massive commercial success. GoPro's big innovation wasn't necessarily strictly the cameras themselves, though. Rather, it's the line of mounting accessories built around the products that were designed for specific purposes such as surfing, motorsports, helmets, and my personal favorite, a dog harness. Today, GoPros are ubiquitous (Figure 5-3), and they've transformed our relationship with video cameras completely. Before GoPro, camcorders were big, delicate, and expensive pieces of equipment that were cumbersome in terms of capturing video, to say nothing of later using the video you've captured (remember DV tapes/decks?). With a complete end-to-end product that starts with effortless capture and ends with publishing, GoPro has been massively influential on how we capture and consume video.

FIGURE 5-2
GoPro HERO 28mm camera

FIGURE 5-3
GoPro HERO3 and Session

WHAT A DIGITAL CAMERA DOES

Before we go any further in terms of talking about design for wearable cameras, it's important to get our heads around what cameras actually do and the data that they produce. On a high level, digital cameras simply focus light onto an image sensor that stores a single color value for each pixel. Most digital cameras use what's called a CMOS (complementary metal-oxide semiconductor) sensor. These sensors are tightly packed grids of individual sensors that convert light into current. They are called photodiodes, and they record the brightness of light on a scale of 0 to 255, 0 being no light 255 being the brightest light they can sense. To record the color of the light, each sensor has a red, green, or blue filter over it allowing only that wavelength to pass through. To construct a single pixel in the camera's memory, the processor in the camera uses a 3 × 3 grid of photodiodes. The center photodiode is green filtered, and additional information is added from the surrounding red, green, and blue photodiodes.

The data that creates an image is simply a list of the color values that are captured by the CMOS sensor. Each image is composed of a total number of pixels, which is derived from the number of pixels that make up the width of the image multiplied by the number of pixels that make up the height. The value for the width lets the device displaying the image know when to start a new row of pixels. For example, take a look at Figure 5-4, which shows the picture that I took using my first-generation Narrative Clip. The total number of pixels that make up the image is 5,038,848, or 5 megapixels. The image is 2,592 pixels wide by 1,944 pixels tall. The data that makes up the image is a list of 5,038,848 individual color values, and those color values are displayed in 1,944 rows of pixels that are 2,592 pixels wide. Zooming in to the upper-right corner, we can see a 25 × 25–pixel segment of the image. The first pixel on the list is in the upper-left corner, and the 54,438th pixel in the list is the 6th pixel in from the left, in the 22nd row of the image. Zooming in even further to a 5 × 5–pixel segment of the image, we can look at the individual color values for the pixels. For instance, in this particular image, the first pixel has a red value of 191, a green value of 208, and a blue value of 233. These RGB color values are the same color values that you use to make colors for any form of digital media. If you drag your cursor around the color picker in Photoshop, you'll see how all these values combine to create colors.

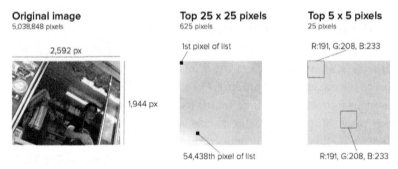

Original image
5,038,848 pixels

2,592 px

1,944 px

Top 25 x 25 pixels
625 pixels

1st pixel of list

54,438th pixel of list

Top 5 x 5 pixels
25 pixels

R:191, G:208, B:233

R:191, G:208, B:233

FIGURE 5-4
Pixel positions and color codes

Surprisingly, how this data is stored is also very important to design. In the binary digital world, bits are how information is stored. Because its binary, each bit is a 1 or a 0. If you have two bits, there are four possible combinations: 00, 01, 10, and 11. If you have three bits, there are nine possible combinations. Four bits have 16, and so on. The reason our RGB color ranges are from 0 to 255 is because each color channel is 16 bits (16^2 = 256, and 0 counts as one of those states), and this means that each individual pixel is 24 bits of information. Meaning that the sensor that records 5,038,848 individual pixels at 24 bits per pixel takes an image that in its raw state is more than 15 MB. But, when I look at the size of the file for this image from my Narrative Clip, it's only 652 kilobytes? How can this be? It's because of compression. If we look at a different part of the image that we were talking about before (Figure 5-4), you can see that the pixels seem to be forming these weird 8 × 8 blocks. These are an artifact of compression called *macroblocks,* where the color values are combined if there's not a lot of detail. There's a whole lot of math that happens there that I don't understand, but that's basically the difference between shooting raw images and shooting JPEG images. It also explains why simpler images have a smaller file size than more complex images.

Digital video cameras work pretty much the exact same way as still cameras except they're doing it multiple times per second, which makes compression even more important. If we were shooting full HD (1,920 × 1,080) at 30 frames per second (FPS) without any compression, we'd be looking at about 51 gigabytes for a 5-minute video (not including audio or anything else that comes along with it), which would severely limit the usefulness of any video camera. Video cameras use two types of compression: *spatial compression,* which is similar to the still images grouping similar colors, and *temporal compression,* which spans multiple frames and takes advantage of redundancy from frame to frame. Basically, if a pixel doesn't change, it doesn't need to rewrite the pixel in the next frame; it will just say that it's the same as the previous frame. The actual frames that are stored in an image sequence for an MPEG video are *I-frames,* which are basically JPEGs; *P-frames,* which mostly hold instructions that refer to the I-frames and are about half the size; and *B-frames,* which fill in the gaps with algorithms that predict the transitions between the I-frames and the P-frames and are about a quarter of the size of I-frames.

The last piece of the puzzle here is one of the most important: *metadata.* Metadata is stored in the Exchangeable Image File (EXIF) format and is hugely valuable if you want to do anything at all with the images after you've captured them. EXIF data ranges from information like the shutter speed to the GPS data indicating where the image was captured. For example, if you look at the EXIF data from a picture taken by a camera with many other sensors on it, such as an iPhone, you can see some interesting things about the photograph, like ISO speed, pixel resolution, metering, and color space. But you also can extract a lot of other information, such as the exact time it was taken, the latitude, longitude, and altitude, as well what direction the camera was facing when the photo was taken. I can type this information into Google Maps and zoom to right where I was standing and exactly what I was taking a picture of when I took that shot!

Handling the Data

So now you're asking yourself, "How could this possibly be important to me?" The fun answer to this question is in the section "What's Possible in The Future," later in this chapter. The most pragmatic answer is that you're going to need to deal with the amount of data stored on your device, servers, or hard drives. Not only is this data costly to store, but it's also difficult for a user to sort through this massive amount of information without any kind of assistance. For example, if I wanted to delete every image that's completely black because I was wearing my camera at night, I can just find the ones for which every pixel value is 0 and then drop them. If I want to edit the video of me going off a "sick ski jump" but I had only my phone, that video will be highly compressed. As it turns out, one of the biggest challenges with producing this much data is actually getting any value out of this much data.

NARRATIVE'S MOMENTS

Narrative's data challenges are huge. With the Narrative Clip 2's larger image sensor shooting a picture every 20 seconds along with some short videos, you're pushing about 2.5 GB to the company's cloud every day, and without compression, that would be almost a half terabyte. But image compression isn't the difficult part: what's tricky is somehow taking about 2,000 pictures a day and making them not only manageable, but meaningful to the people looking through them. Comparing the images pulled directly from the camera and the images that make it to Narrative's online service says a lot about the processing it does on its servers. First, all black images are filtered out; those pictures obviously won't mean much to you, and there could be a bunch of them if the camera gets covered up or you're walking around in the dark. The second thing to go are blurry pictures, which aren't that hard to filter out by looking at the patterns in the pixel values. Generally, pictures that are sharper, meaning that they are more in focus, have higher overall contrast.

So, after you get the junk out, you still have a lot of pictures. From here, Narrative's service groups pictures in to moments (Figure 5-5) that are chronologically ordered. A quick glance at the time stamps for the moments show that they can last anywhere from a couple minutes to more than five hours. The moments look like they're divided up like scenes in a movie: when the service recognizes I'm in a consistent setting, that's a moment; when I go from an airplane interior to walking through an airport, something has obviously changed about my surroundings, and so it's a new moment. I don't know exactly how Narrative did this, but if I were to make a supposition, I would say that if you took all of the color values and averaged them, and if that average passed a certain threshold, it's probably be a good idea to start a new moment. The more interesting part is how the company picks the individual pictures to showcase in the timeline before you click into an actual moment. They're pretty representative of the moment, and if I had to guess, they might be the most dynamic pictures in terms of color range, but that's just a blind guess. No matter how the developers pulled it off, some serious work went into filtering, separating, and showcasing the pictures in a way that's meaningful and interesting.

FIGURE 5-5
A moment

GOPRO'S MOBILE VIDEO EDITING

GoPro has a very different problem to solve with handling the massive amounts of content generated by its users. Video is big. The files themselves are big, and getting any real value out of them is a lot more difficult because manipulating the files is a much more complicated. Although it's similar to Narrative's problem of making massive amounts of data useful to its users, GoPro's approach is very different: mobile video editing. Given that these cameras are intended to operate where you don't have easy access to a laptop, to actually use any of the content that you've created, you need to get that content onto your phone and edit together something interesting there. The first step of this process is actually moving video clips to your phone. GoPro has had a solid solution to this problem for a while now, beginning with the Hero2 Professional: GoPro cameras come with a WiFi antenna that not only gives you a very useful live preview from the screenless cameras, but, more important for our purposes, allows you to access the memory card to transfer video clips to your phone or share them directly to certain social media platforms.

GoPro didn't stop at simply transferring short clips to share from your phone. In February 2016, GoPro purchased two mobile video editing applications: Replay (renamed to Quik under the GoPro banner) and Slice. Quik is more of a hands-off video editor that focuses mostly on interesting video effects similar to what you'd see in something like iMovie 10 years ago, for example, putting frames around your video and having weird transitions. Slice, on the other hand, feels a lot more like a contemporary prosumer video editor that focuses on more practical functionality like trimming clips, adding text for titles, and a really great narration feature that lets you record a voiceover as you're playing the video in the editor. After testing out both applications, I'm thoroughly convinced that GoPro has completely solved the end-to-end, high-quality video capture, edit, and publish problem.

Computer-Mediated Reality

Though not very pervasive right now, in terms of wearable devices at least, the future role of wearable cameras will be in their support of *computer-mediated reality*. It's important to distinguish computer-mediated reality from its more commonly known subset, *augmented reality*. Computer-mediated reality is a broad term that describes any type of digital mediation of our perception of our environment. This could involve overlaying contextual information mapped to your field of vision (a form of augmented reality), it could be mapping virtual objects to our environment (*mixed reality*), it could be subtracting things from our perception (*diminished reality*), or it could even simply be improving our natural perceptual abilities. Computer-mediated reality is an incredibly exciting field of work, and I'm certain it will change the way we understand the world in the near future.

Computer-mediated reality relies on a camera for data capture and (this is where the technical stuff we looked at earlier comes in to play) *computer vision*! Computer vision is simply processing visual data from cameras and pulling information out of it. Computer vision seems incredibly complex—and it is—but it's not completely inaccessible if you know what's really going on. In the section "What a Digital Camera Does" earlier in the chapter, we looked at how visual data from cameras is organized. Recall that it's basically a list of color values, so to get information, we just need to understand what certain behavior in the image looks like in the data.

COMPUTER VISION AND SUPPORTING TECHNOLOGIES

For a basic example of computer vision, I'm going to walk through a simple system (Figure 5-6) that I built a few years ago for a project I was working on for which I needed to automatically count the number of people walking into and walking out of a door by using a video camera. In the image in Figure 5-6, you can see what this very simple computer vision looked like: the left side is the processed camera feed, and the image on the right is the raw data from the camera. The first thing I did to the video feed was to run it through a frame-differencing algorithm that pointed out what changed from one frame of the video to the next. The algorithm compared each pixel value of the video to the previous frame's value for the corresponding pixels. If the difference between those two color values exceeded a certain threshold, I would make that pixel show up as white. If it was below that threshold (didn't change enough), it would show up as black. In the image, you can see me walking through the frame, and on the left side I'm just a white silhouette as a result of the frame differencing.

FIGURE 5-6
Basic computer vision system

On top of the frame-differencing algorithm, I overlaid a blob-detection algorithm, which automatically identifies chunks of similar pixels within a single image by comparing the color values of all of the pixels around each pixel. This algorithm ran on each individual frame and compared the current frame blobs to the previous frame blobs. If the current frame blob was roughly the same size and within a certain distance from the previous frame blob, I could safely assume that this is the same object that I can track across the screen. In the image, the box around me is the box that identifies me as a blob, and the dot in the center of that box is over the pixel coordinates from the blob center. Thus, after a blob is identified and tracked across the view of the camera, I put the x- and y-pixel coordinates of that blob into a list, and if the first coordinates of that blob start within a certain range (right around the door) and end outside of that range, it means someone is walking in the door. Conversely, if the coordinates start outside of that range and end inside of that range, it means that they are walking out of the door. The green box on the right side of the processed image marks the threshold for where the blobs start and stop.

In Figure 5-7, I used a form of computer vision to tie a mixed-reality object to a specific Fiducial marker pattern, a predefined symbol that the software recognizes. The program knows what the marker looks like and matches the orientation of the virtual object to the orientation of the physical marker, though more modern mixed-reality technology doesn't need to tie the objects to markers. It's not only computer vision that enables these technologies, though; there are a lot of other sensors and pieces of information. For example, most traditional phone-based, mixed-reality applications like Pokemon Go (Figure 5-8) take advantage of geolocation data, the gyroscope, the accelerometer, and the compass in your phone to understand where it is located and exactly what direction it is pointing.

FIGURE 5-7
Basic mixed-reality example

FIGURE 5-8
My wife, Marita, playing Pokemon Go

AUGMENTED REALITY

Augmented reality (AR) is a subset of mediated reality in which data from your physical environment is used as an input to create additional contextual information that is presented to you in real time. Most popular examples of this technology have this augmenting information visually mapped to the camera input (like Monocle), but it doesn't have to be. A good example of nonmapped AR is the Google Glass application called CamFind (Figure 5-9). CamFind allows you to take a picture of what's in front of the Glass' camera and it runs the image through an image recognition service and then visually displays the data on the Glass' screen and reads the description verbally.

A classic example of AR that has yet to be realized commercially is its potential ability to pull up contextual information in interpersonal settings based on facial recognition and access to social media profiles like Facebook or LinkedIn. AR really gets interesting when we look at how the brain processes visual information. If we put it in terms of bandwidth, our eyes have the ability to sense roughly 10 MBps, yet our minds only have the ability to process 40 bits of this information per second. If we can get computers small enough to comfortably wear for extended periods of time that can capture and process this visual information faster than our brains can do these things, we open up a lot of opportunities for increasing our understanding of what's happening around us.

FIGURE 5-9
Google Glass identifying my dog

MIXED REALITY

Mixed reality (MR) is when three-dimensional objects are virtually inserted into our field of vision to create the perception in our minds that they're actually there. MR is likely going to be one of the biggest shifts in our understanding of technology and computing as a whole, but as of now, there aren't many consumer products that support it. There are a handful of very basic MR applications for Google Cardboard, but nothing that goes beyond what feels like a demo. The main problem is that there aren't any consumer platforms that support the technology, this will change soon though. For the moment, there are three major players that are releasing MR headsets in the near future. Microsoft has released a $3,000 developer edition of its Hololens MR system; Kickstarter-backed Meta has shipped a developer edition and is taking preorders for a consumer version; and the secretive startup Magic Leap (Figure 5-10) is developing very advanced MR technology.

FIGURE 5-10
MR example from Magic Leap

So what will MR do? Well, in the beginning it is going to be limited to the content that's available, which is overwhelmingly two dimensional. Nearly all of the digital content we have currently is built for two-dimensional screens. This isn't all that bad though. Early demonstrations of Microsoft's Hololens show a cool use case in which you can basically have screens wherever you want in your space, at whatever size you want them. After the content evolves a bit, video games and adult entertainment will obviously be major innovators in the space, but there's also three-dimensional movies, sports, and modeling.

DIMINISHED REALITY

Diminished reality (DR) is an interesting concept by which, through certain of these mediated realities, things can and should not be available to us, and are therefore blocked from our perception. My favorite fictional example is from the British science-fiction television show *Black Mirror*, in the episode "White Christmas," everyone wore permeant AR contact lenses, and if you didn't want to see someone, you could block them from ever seeing you again. Later in the episode the protagonist receives a criminal punishment of being both blocked by everyone and not being able to see anyone either (Figure 5-11), which is nightmarish.

FIGURE 5-11
Diminished reality example from the TV show Black Mirror

In the present world, however, there are more immediate concerns that are addressed by diminished reality. With technologies like Google Street View (Figure 5-12), Google Earth's satellite images, and the potential ubiquity of wearable cameras that come along with the commercial rise of AR and MR, there are a growing list of things that people don't want recorded. A great example of this is the algorithmic blurring of both faces and license plates that we see regularly in photos and videos, with the option of requesting certain items be manually blurred further. In 2014, to address this issue with people constantly recording and internet-connected cameras, a group from Indiana University developed a software prototype of what it calls Place Avoider. Place Avoider blocks specific sensitive content based on image recognition that if flagged by its users—for instance, if you don't want pictures taken in a certain room of your office building—you would photograph the room and flag the images so that when similar images showed up on the networked recording devices, they'd be blocked from the system.

FIGURE 5-12
Diminished-reality example from Google Street View

THE FUTURE ON (OR IN) YOUR EYEBALLS

Cameras embedded in contact lenses or into our eyes themselves sound about as absurd as iPhones would have been 20 years ago, but believe it or not, they're coming. 2016 saw two very interesting patents awarded in this area from Google (Alphabet) and Sony.[1] Sony's patent is a camera contact lens (Figure 5-13) that you've seen in movies like *Mission Impossible*, and it's going to be thrilling if they ever develop the technology.

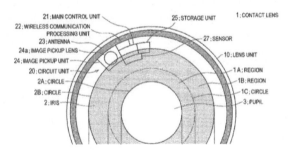

FIGURE 5-13
Sony's contact lens

---

1 Yoichiro Sako, Masanori Iwasaki, Kazunori Hayashi, Takayasu Kon, Takatoshi Nakamura, Tomoya Onuma, and Akira Tange. Contact Lens and Storage Medium. US Patent US 2016/0097940 A1, filed February 12, 2014, and issued April 7, 2016.

The more interesting patent comes from Alphabet, or, more specifically, Verily Life Sciences, which is for an always-on computer/camera (Figure 5-14) that's actually designed to be surgically implanted onto the eyeball and fuse with our natural lenses.[2]

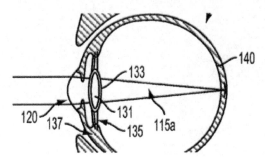

FIGURE 5-14
Alphabet's implant

Pulling It All Together

As technology shrinks in size and grows exponentially in power, wearable cameras are evolving into very interesting devices that have serious potential to change the way we interact with technology as well as the way we perceive our world. In the next couple years, we'll see amazing advances in computer-mediated reality devices introduced to the consumer market, and after that we'll see even more wild computers that look nothing like the computers we know today. With these advancements, there are massive risks around personal privacy and interpersonal relationships that we'll all need to navigate together, and with the right design, they might be able to be avoided altogether.

---

2 Andrew Jason Conrad. Intra-ocular Device. US Patent US 2016/0113760 A1, filed October 24, 2014, and issued April 28, 2016.

[6]

Cognitive Wearables

Cognitive wearables are devices that either measure or affect our cognition. They are a part of an emerging and powerful field of technology that is clearly among the most exciting. From a design perspective, there is an incredible opportunity to help people better understand themselves and their lives, and an equally massive risk around mischaracterizing information about something so intimate. This chapter covers the landscape of what's out there right now in the cognitive space, and the unique opportunities and challenges that come with communicating information about the subconscious mind to the conscious mind.

Fitness Trackers for Your Mind

When I first began hearing about what I now refer to as *cognitive wearables*, I had a difficult time describing them. I went through a bunch of different names like psychological wearables and mental wearables, but after a while I began calling them subjective wearables because it seemed like what they were measuring or affecting seemed to be subjective. After talking to a few of the people developing the technology and asking them about specific terminology, I began hearing the same thing: these devices are not subjective. Things that I had previously understood to be dependent on personal interpretation, such as stress levels, concentration, and affect, turn out to be not very subjective at all when it comes to measurement. As it turns out, the things that we feel in our minds have very real physical symptoms, and that's what these devices do—they sense these physical signals of mental activity and convert them into information very similar to what fitness trackers do with our movement.

There are a lot of similarities between some fitness trackers and cognitive wearables. You could say that many of the cognitive devices on the market right now are fitness trackers for your mind. Both categories of wearables rely on sensing *biological indicators* (bioindicators) which are symptoms of other things going on. In the case of cognitive wearables, those indicators are things like breathing patterns, skin perspiration, heart rate, and the electricity that's produced by brain activity. Most cognitive wearables and fitness trackers share several overall goals, which are to tell you something about yourself that you didn't know before; thus, they run into the same problem of losing their usefulness over time as that information becomes repetitive.

What makes designing around cognition both interesting and dangerous is the fidelity and accessibility of the information that's reported. If you look down at your pedometer and you know that you've taken about 300 steps but the device shows only 50, you can dismiss that because the data that your device is reporting back to you has a physical analog that is very easy to understand. Everyone knows what a step is, and if the pedometer is off by a couple hundred steps, it will be evident. Cognitive wearables don't have the luxury of an easily understood analog. We don't have commonly understood units for stress or happiness, and this problem complicates the design in two major ways: developing the technology, and then communicating the data back to people.

If I'm developing a pedometer, I can walk around my studio and manually count the number of steps that I've taken and then compare it to the data to see if my algorithm is correct; it's pretty straightforward. For cognitive devices, though, how can I be sure that what I'm feeling right now is stress? How can I know that I'm happy? The biggest hurdle in developing these devices is taking something as inaccessible as the human consciousness and measuring it objectively, which is definitely possible—it's just not easy. When you do have a handle on the bioindicators and you're confident you're getting meaningful information, how do you then communicate that to people? Are you medium-sad or just bored? Is your arousal because of anger or happiness? How do you make this information meaningful and useful for people experiencing these things?

Stress and Focus

The current generation of consumer cognitive wearables center around reporting on two things: stress and focus. These two mental states are ideal for early trackers for two reasons. The first is that they have the most pronounced physical symptoms. If we look at our physical manifestation of stress, it's pretty obvious: our heart rate goes up, we sweat more, we breathe differently, and specific frequencies of the electrical activity in our brains are very clearly more pronounced. Other cognitive processes such as happiness can be a lot more complex to both sense and interpret. The second reason stress and focus are leading the way to cognitive wearables is that they're relatively harmless to communicate to someone. There's a lot of risk involved in telling people that they're displaying symptoms of sadness as opposed to stress and focus.

SPIRE

Spire is the first wearable that I am aware of that attempted to measure cognition. Oddly enough I found out about it from my tech-savvy mother-in-law, who wears Spire religiously and swears that it was accurate and was genuinely helping her calm down. Spire is a beautiful pager-style device (Figure 6-1) that you wear on your undergarments. It detects the movement of your torso to measure your breathing patterns. The product is sold as a means to identify stress and take action based on those measurements. The product comprises two parts. First, there's the monitoring/reporting, through which your status is reported as tense, calm, focused, or active (when you're walking around). The second part is what Spire calls Boosts, which are short audio-based interventions that are meant to affect your status. For example, when you're tense, you're prompted to do a Calm Boost to relax a little bit.

Our respiratory patterns are incredibly rich pieces of feedback that reveal much about our cognitive state. It's not just the frequency of breaths, there's also the time it takes to inhale, exhale, and general respiratory consistency that are all reliable indicators. Frequency is definitely the major indicator. It makes sense: if your breath is quick, you're in a state of arousal and you're probably not relaxed, but that doesn't necessarily mean you're stressed out. If your breathing is faster and consistent, the Spire reports that you're focused. This moves to tension when your breaths are quicker and more erratic. There's plenty of science to back this up, but it just kind of makes sense logically that these patterns would indicate stress and focus. Conversely, if your breathing is slower and consistent, you're calm.

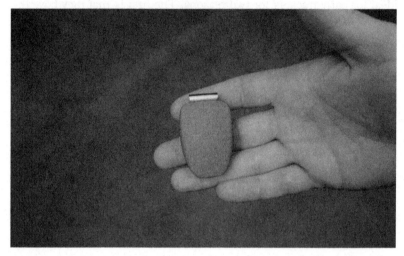

FIGURE 6-1
Spire

FIGURE 6-2
Spire screenshots

So, the big question with new things like this is this: does it work? I was wearing Spire during an unusually stressful part of my life. My sister had recently been diagnosed with cancer, and I was more than 2,000 miles away in California. At any moment, I could get a phone call that would be absolutely devastating, and every single time my mom called me, I immediately got a notification from Spire saying that I was tense. This continued for a long time until my sister's outlook became consistently better. My phone calls with my mom slowly began to focus less and less on cancer. Accordingly, I no longer received the notifications from Spire that I was stressed out when she called. It also seems pretty consistent in the other states, as well; for example, at work I'm constantly switching between focus and tension depending on what I'm doing, which seems to match my productivity and how well things are going from moment to moment.

So it's accurate. But is it useful, and what do you get out of it? Spire is about as useful as any fitness tracker, and it runs into the same issues as fitness trackers in terms of longevity of use. The Spire uncovers things that stress you out in your life that you might not have known about before. My mother-in-law learned that phone calls and driving often stress her out, and with that knowledge she was able to be proactive in terms of mediating that stress. However, my mother-in-law doesn't wear her Spire that much anymore because she figured out what stressed her out already. After a certain point, she was able to understand the patterns of stress, and after that pattern was obvious and consistent enough, the repetitive data stopped becoming useful.

MUSE

Muse is an electroencephalography (EEG) band that's worn across the forehead and monitors electrical signals from your brain. Muse is marketed and sold as a meditation assistant. You are instructed to put on the headband and go through a three-minute guided meditation session. Then, the device gives you feedback on how well you're doing at meditating. The feedback mechanism is pretty interesting: when you're meditating, your eyes are closed, so it's all audio. You hear the relaxing sound of small ocean waves calmly crashing on a beach, and when your brain waves are more active, the wind on your beach grows louder to match the activity in your mind. When your mind is more clear, the wind is calm.

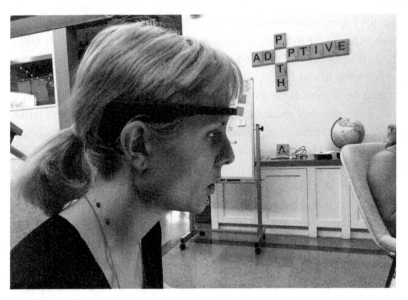

Using Muse is a very interesting experience. The first time you take it out of the box, you go through a calibration process in which you're prompted to list things in your head, like animals, countries, and toys. This calibration is how Muse establishes levels on your brain activity to use as a baseline when you're attempting not to think about anything during meditation. It doesn't sound all that weird, but it was the first time that I had to be accountable for what was happening in my head. It was the first time that barrier of privacy was breached and I couldn't just think about whatever I wanted. It's the first time that my thinking about something was acknowledged as a physical process and objectively measured!

As a consumer product, Muse definitely does its job. The Muse consumer application (Figure 6-4) is really well designed and gives you great feedback during meditation, and afterward you get a report about your brain activity. You're encouraged to meditate for at least 10 minutes each week and are given feedback about your progression over time. After using it a few times, I really did feel more calm, and of all the cognitive wearables I've experienced, it was the most effective at affecting my cognition. I probably wouldn't have tried meditation without the device, and I definitely wouldn't have succeeded at meditation in any way without it.

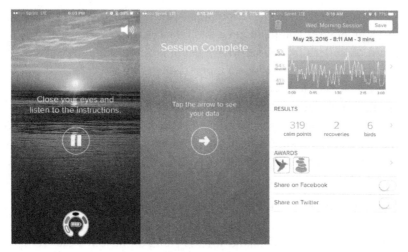

FIGURE 6-4

Screenshots from Muse iPhone app

Even though it's great as a consumer product, Muse really shows its power when you dig into its developer kit. Muse has a huge presence in the academic research community because it's relatively inexpensive and very easy to set up. After downloading its research tools, I was able to get my raw EEG scores up on my screen in about 10 minutes (Figure 6-5). It's super easy and also really interesting to get a live readout of what's going on in my head!

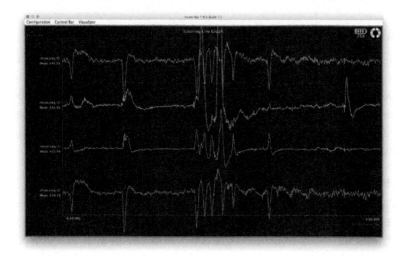

FIGURE 6-5

EEG readouts from Muse

When you first see it, the information can be overwhelming, there's just a lot of stuff going on and a lot of data that you've probably never heard of, so let's talk about what's going on here. The EEG data is a measurement of voltage fluctuations of the electrical activity in your brain. The electrical signals are divided up in to different bands based on their frequency and are named after the Greek letters alpha, beta, delta, gamma, and theta. Each bandwidth is associated with certain mental processes; for example, activity in the beta range is associated with focus and anxiety, and the alpha bandwidth is more active when you're relaxed.

I got to talk to Muse founder Ariel Garten, and I specifically asked her about how you would develop software that could accurately interoperate EEG signals to understand what is going on in your head. Here's how she explained it to me:

> If you want to build an algorithm for kitten reaction, you need to show a couple thousand people with kittens and get that reaction, and get it well pegged and then come up with your algorithm from it.

It's basically a mass data gathering and classification process. There are certain things that are less nuanced than others, though. For example, focus and relaxation are pretty obvious, whereas affect might take a little more work. Either way, if you're looking to develop a cognitive wearable, going straight to the brain activity isn't a bad place to begin.

Altering Consciousness

Beyond measuring our consciousness, there are also efforts to affect our consciousness similarly to psychoactive medications, except using wearable devices. This general concept has existed for a while in the form of electroshock therapy, where electric currents cause a seizure in the brain in order to alter the brain's chemistry, the emerging consumer wearables are much less destructive.

THYNC

Thync is a cognitive wearable that's quite different from Spire and Muse. It doesn't measure anything. In fact, it does the exact opposite of Muse: it sends electrical signals *to* your brain. Thync is a curved, triangle-shaped, Bluetooth-connected device (Figure 6-6) that snaps in to two different types of electrode strips, one that's meant to calm you, and the other that's meant to increase your energy. To use the Thync, you snap the device to an electrode strip, attach the electrodes to your forehead and neck, and use its smartphone application to control your "vibe," which is what they call the electricity sessions. The vibes are said to target and stimulate your sympathetic and parasympathetic nervous systems, respectively. The "energy" mode of the experience stimulates your sympathetic nervous system, which is responsible for our subconscious "fight or flight" reactions, and the "calm" mode is targeted to your parasympathetic nervous system, which is associated with our "rest and digest" activity.

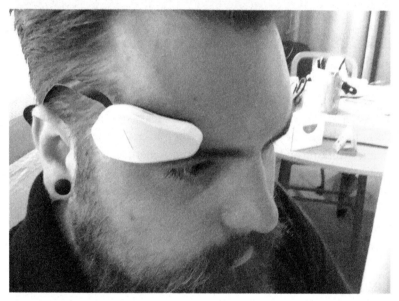

FIGURE 6-6
Wearing the Thync device

When you first use Thync, you see an instructional video showing you how to position the device on your head, and you select which type of electrode strip you want to use. You then place the electrodes on your head and begin one of its introductory sessions, which are a 15-minute guided meditation-like exercise during which it teaches you how to calibrate the device and discover how comfortable you are with different levels of electricity (Figure 6-7). Initially, this is a really weird experience. It starts out pretty low, but you soon begin to feel a little bit of tingling on your skin between the electrodes, and if you turn it up too much you can actually begin to feel some discomfort.

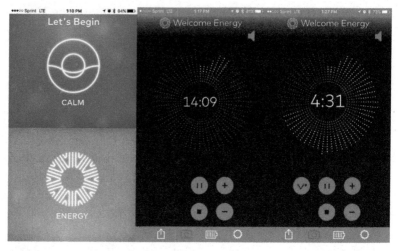

FIGURE 6-7
Screenshots from the Thync iPhone app

Does the Thync work? Does it actually increase your energy? I think it does, but I can't be certain because the in-session coaching suggests that it's working and you should feel more energized, see objects more clearly, and perceive your immediate surroundings as "brighter." I definitely feel *something*, but it might just be the general sensation and novelty of zapping my brain with electricity. When I ask my friends and family to test out the device, some of them feel the effects more than others, but for the most part people did claim to feel more energy.

Learning to Learn

The same thing that makes cognitive wearables so interesting and useful also makes them particularly difficult to design for: they're giving us information that we've never had access to before, and because we've never been exposed to this type of information, it's tough to know how to best communicate it. The difficulties around communicating this information lie in three areas: the language isn't very established; it's difficult to connect cognition to the events or circumstances that contribute to those cognitive processes; and, quite frankly, people might not be ready to know some of the things that they could potentially learn by using devices that measure their cognition.

LANGUAGE

Fitness trackers have an established language that's based on the obvious physical actions that they're attempting to measure—steps, calories burned, and hours slept are all relatively well understood units of measure because they correspond to commonly understood parts of our lives. Cognitive wearables don't have that luxury. Right now, we have very broad terms like focus, tension, and calm that are mostly understood, but as cognitive devices advance, it's going to become significantly more difficult to describe more specific aspects of our cognition. We currently don't have units for excitement, sadness, boredom, or elation. Thus, we can definitely say things like, "You were a little sad for 45 minutes today," but what does "a little" mean in that sentence? Could standardized units like "700 sads" work better? Or possibly sentences like, 75% sad, 15% anxious, and 10% frustrated? Further compounding these problems is the subjectivity of understanding our feelings. Whereas the biological indicators that we measure with these devices might be completely objective, what those indicators mean can be wildly different from one person to the next. For example, what I might interpret as moderate elation might be low-level happiness for the next person.

CAUSALITY

The next big hurdle for communication to get over is trying to provide some type of causality. Similar to fitness trackers, the raw descriptive data is only useful to a certain point. The true power of the devices comes from their ability to provide more prescriptive information. It's much more difficult to be prescriptive when you need to connect cognitive states to what could be influencing them, and for now, at least, it looks like those dots are going to have to be connected by the users themselves.

A lot of this comes down to frequency of communication; for example, Spire works really well for flagging what stresses you out because when you're tense, it will shoot you an alert. After you get a couple of those alerts, you can begin thinking back and looking for patterns, such as driving or talking on the phone. On a longer timeline, and with more emotion-centric data, you might notice that after a certain point in your life, you've been feeling differently, maybe you look back at a certain date on your timeline and notice that since that point you're significantly happier and try to figure out what happened around that time that might have contributed to your happiness.

INTIMACY

The last big part of the learning process for cognitive wearables is the intimacy of the information. Like I said earlier, this type of information has never been consciously available to most people, and quite frankly, it might be tough to see objective measurements of things that we might otherwise try to ignore. On top of that, externalizing that type of information makes it potentially available to other people. Suppose that you're really stressed out, but you're trying to maintain an air of calmness. People around you could potentially see that you're really stressed out without you acting any differently.

When I first got the Muse, I was playing around with the developer tools in our studio, I had my raw EEG data streaming on my computer monitor as I was working on some other things. My friend Iran walked by and we talked for a moment. When she was about to walk away, she remarked that it was weird talking to me while she could see my EEG data changing so much depending on what we were talking about. Suddenly, I felt really embarrassed that she could see my mental activity! This very intimate thing—my brain waves—were exposed, and if I were feeling a certain way about any given situation, I couldn't hide it.

FIGURE 6-8
Wearing the Cognitive Accountability System at Adaptive Path

I decided to take this concept a little further, so I put a simple live stream of my brain activity on my phone and wore it around my neck on a lanyard. I coached the people around me to interpret the information; for example, if the blue line was down, I probably wasn't paying attention to what was going on. If the red line was up, I was more engaged, and if the green and yellow lines were higher, I was probably more excited about what was going on. I called it my cognitive accountability system and wore it around the studio for a couple days. It was really interesting to be in a meeting and begin losing focus and have one of my coworkers point at my phone and say, "Pay attention!"

Pulling It All Together

Cognitive wearables are coming, and they're going to change the way we understand ourselves. With normal activity trackers, we're only getting a small sliver of our overall state of being. The addition of the cognitive pieces of that puzzle will give us a much more complete idea of what's going on in our lives. Combining as much data as possible is the key here, looking at every factor that could possibly contribute to different mental states, such as exercise, location, social interactions, and other circumstances. Discovering and externalizing the correlation of multiple factors of available information will give us access to an understanding and control over our minds and bodies that we've never had before.

[7]

Service Design

Everyone's design process is different, and even individual processes are different from project to project. Add to that all of the wildly different types of wearable devices and services, and you can see how it wouldn't make any sense to get too prescriptive about the details of design. So rather than lay out step-by-step tutorials, in this chapter I discuss my general design process and some specific tools that are useful to communicate those designs as well as some of the future-focused techniques and ideas around machine learning that I'm working on.

Service Design

What precisely is *service design*? Simply put, service design uses design methods and approaches to craft a holistic, end-to-end experience of a service: how its operations and processes unfold, and how value will be delivered over time to its internal and external users. This is different from product design or user experience (UX) design because in traditional UX, the design decisions that you're making usually focus on a single touchpoint, even if those decisions are made with an awareness of the greater context. In practice, service design usually takes a higher-level approach to the end-to-end process of delivering a complete service. If we look at the design of a hotel experience, service design would involve mapping the end-to-end experience, considering the design of visible touchpoints like the registration desk or online booking tools, the specific role and experience of the person working at that desk, and the internal systems that employees rely on to register guests. In contrast, UX or product design might take a more in-depth look at any one of those components.

The benefit of thinking about the design of wearable devices as a service as opposed to a product stems from the fact that the value derived from contemporary wearable devices comes from what the device enables as opposed to the device itself. In his book, *Smart Things*, Mike Kuniavsky refers to these devices as "service avatars":[1]

> When information produces most of the value for users, the hardware and software through which it is experienced takes on a secondary role. Rather than being the center of attention, it becomes a conduit for delivering information-driven services.

Mike goes on to explain that the device itself is only an entry point to the service, and they're avatars because to the user they act as a physical symbol of the service, a physical projection of an abstract concept.

Apple's Healthkit, a central hub for health and activity data, is a great example of this type of relationship. When I wear my Jawbone UP, it sends my activity information to the service. If I'm wearing my Apple Watch, instead, it reports that same information, or if I'm wearing neither, my phone itself can act as the source of data. The devices that record the data act only as an entry point to the larger service that is the remote logging and analysis of the information over time.

DISNEY'S MAGICBAND

The biggest wearable service design project that I can think of is Disney's billion-dollar MagicBand. The device itself is simply a strap with an RFID tag (Figure 7-1), a radio transmitter with a 40-foot range and a two-year battery life, but its impact on the overall experience of the park is massive. In theory, it's not much different than the key fob that you might use to enter your office every morning, and it definitely does that kind of thing, but there's a lot going on behind the scenes.

The MagicBand experience begins when you buy your park admission tickets. You fill out a questionnaire, register everyone in your group, tell Disney what you're interested in, and make reservations at restaurants, attractions, and character greetings. From here, Disney spits out a detailed itinerary that allows you to do everything you want to do with

---

1 Mike Kuniavsky, *Smart Things: Ubiquitous Computing User Experience Design* (Morgan Kaufmann, 2010).

your limited time in the park while simultaneously giving Disney the means to exert some level of control over the flow of people through the park.

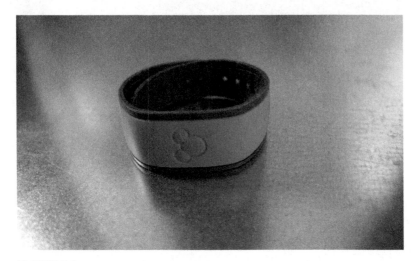

FIGURE 7-1
The MagicBand

When you arrive at the park, the MagicBand is your seamless key to everything. It checks you in to your Disney Resort (Figure 7-2),[2] it's your room key, your park pass, your credit card at restaurants and shops in the park, and your ticket to all of the special events you scheduled a month ago when you signed up. If you have the My Disney Experience app on your phone—which of course you do—you're guided through the park with a dynamic map that shows you wait times for nearby attractions and real-time instructions to execute your perfect plans for your visit. It even magically displays the photographs that the roving Disney PhotoPass photographers snap of you, minutes after they take the picture.

The experience of visiting a Disney park is now highly influenced by the MagicBand, but the overall value of the experience came from the end-to-end service. In that respect, the device serves as both a symbol of that service and an entry point.

---

2 Chad Erickson, "Epcot - Going Green," November 6, 2014 (*https://flic.kr/p/raSZvn*).

FIGURE 7-2
Checking in with the MagicBand

The Double Diamond

So how do we make this happen? The scope of service design projects is different than what you might see in product design, but in the beginning, at least, the processes are pretty similar. I've been with Adaptive Path for a couple years now and am a firm believer in the *double-diamond* design process, which you can see illustrated in Figure 7-3.

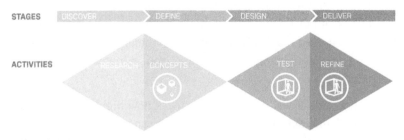

FIGURE 7-3
Double-diamond diagram

The double-diamond model reads from left to right and describes taking a project from the discovery phase through delivery. It's not meant to be taken as gospel because every project is very different, but on a high level, this is how they tend to go:

Research

The left side of the left diamond describes research and basic information gathering. Because this is an iterative process, the first time you go through an iteration cycle, research represents generative research, which is more open ended and not very guided. The next iteration will be based around validation of the low-fidelity concepts you've generated. The third iteration will be more specific around usability:

- Get everything you can out of stakeholders and subject matter experts around what they're trying to accomplish.

- Do an audit of the existing product/service space.

- Identify and interview people who fit your target audience. Figure out what could help them and what motivates them.

Synthesis

After every period of iteration, there needs to be a period of synthesis. This is when you'll perform tasks like affinity mapping and group like pieces of feedback to try to construct a clearer picture of feedback spaces:

- Group pieces of feedback by category or any other way that makes sense for the specific information that you've gathered.

- Define the problem that you're designing for and overall high-level intentions of the potential solution.

- Define touchpoints or physical restrictions around which you're designing.

- Define any principals or core values toward which you want to design.

Concepting/ideation

After you have a good idea of what you want to design and any guiding principles, there's a period of ideation. This generally involves your design team as well as any stakeholders:

- Define parameters or some type of guidance or organization system for sketching, such as specific issues (from synthesis) that concepts should address or different touchpoints or processes that the service will involve.

- Sketch! Create as many concepts as humanly possible in the amount of time that you have and organize them by the issues and touchpoints or service phases you've defined.

- Narrow the concepts. This is usually done by some sort of ranking system or voting system with the stakeholders.

Refine

Take the rough sketches from the concepting process and string them together into a cohesive service. If the concepts involve a digital aspect, you'll bump up the fidelity to make them as realistic as is reasonable for the amount of time that you have. If they're hardware based, build a quick prototype; if they're process based, sketch those to the highest fidelity possible. The purpose of this phase is to produce the most realistic analog possible of your design order to get it in front of people for feedback and validation:

- Make your concepts as real as possible within the given amount of time, through static "look and feel" models, interactive digital prototypes (Figure 7-4), representative images, or faking it on an Android phone (see Chapter 9).

- Identify what information you need in order to verify your concepts.

After the refine stage, you rinse and repeat. You have a prototype that you take back to the first diamond and carry out research. You then synthesize that research and update your prototype to put in front of people yet again (time permitting). The reason for this design process is to learn as much about how people will use and interact with your service as possible without going through the costly process of actually creating the full-scale service and hardware/software/hiring that's needed to support that service. You're figuring out the *what* before you figure out the *how*.

FIGURE 7-4
A prototype for a watch I'm working on

Communication Tools

With so many moving parts involved, you're probably going to have a lot of people contributing to the project, and those people doing very different things. The most prominent communication tools are the *service blueprint* and the *customer journey map*, both of which describe the service in a time-based, map-like manner, but they have very different purposes and information.

CUSTOMER JOURNEY MAP

Journey maps are models that describe the customer's end-to-end experience of the service across multiple touchpoints. The map in Figure 7-5 is for Rail Europe, a US distributor of European train tickets. It describes the entire process of rail travel starting with the customer planning a trip and ending with the post-travel experience. The columns of the experience map describe different phases in the journey, organized chronologically. In this case, they're Research & Planning, Shopping, Booking, Post-Booking, Pre-Travel, Travel, and Post-Travel.

The rows of experience maps are different for every project, but the Rail Europe map contains some pretty common ones: Doing, which is what are they trying to do during each stage (along with the touchpoints they use when they're doing what they're doing); Thinking, which is what are they thinking at each stage; and Feeling, which is what they're feeling at each stage. Other common journey map rows are their emotional experience (observed from research), their expectations of each stage, and key moments.

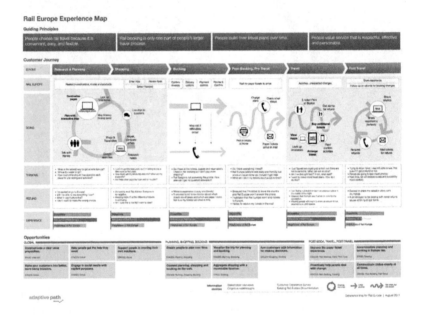

FIGURE 7-5
Rail Europe Experience Map by Adaptive Path

SERVICE BLUEPRINT

A service blueprint (Figure 7-6) is similar to a customer journey map in that it's chronological and a map, but the purpose of the blueprint and the information contained in the blueprint are very different. A service blueprint describes the complex systems that support a service for the purpose of verifying and implementing that service; it's a much more technical document than a journey map.

The main structure of a service blueprint is horizontally divided in half with terminology borrowed from live performances. The top half is the *frontstage*—what the customer sees—and the bottom half is *backstage*—the actions that are not exposed to the customer but support what's going on frontstage. The rows that make up the frontstage half are the customer actions, such as "Customer is greeted by host," touchpoints that the customer uses to interact with the service, such as "customer greeting," and the visible staff that supports the service, the "host." The backstage rows are the backstage staff, the people who support the customer but with whom the customer does not directly interact. These are the processes that need to happen, like "Back Waiter tells Host that table is ready," and the systems that those processes live on.

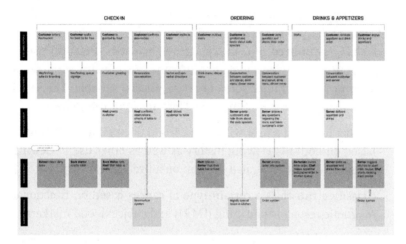

FIGURE 7-6
Example of a service blueprint for a restaurant, (Nick Remis Adaptive Path)

GOING BIG

The preceding examples involve traditional services like travel and restaurants. They're good examples, but this isn't exactly what we're talking about here with wearables. Nonetheless, the same ideas apply for digital services. The main reason I talk about service design instead of product design is service design's emphasis on providing value over a longer period of time and the emphasis of that ongoing value as opposed to the individual product; however, one of the major advantages that wearable devices have in the service ecosystem is the volume of information that they collect! You can do a lot with this data, and the major advances in digital services are going to be the product of machine learning.

Service Design and Machine Learning

Wearable devices, by design, are physically attached to our bodies and are constantly recording information about where we are, how we're moving, our mental state, our sleep, our hearts, and what's going on in our immediate environment. They're incredibly intimate and engineered to collect more information than the human mind is capable of making sense of on our own. To make any use of this massive amount of data, we're going to have to rely on *machine learning*.

Machine learning refers to the field that is concerned with extracting important patterns and trends in data by using computers to analyze that data. A large majority of machine learning techniques rely on methodologies utilized in data science and statistics. Practical use of these algorithms is typically motivated by the goal of using some set of *predictors* or *inputs* to infer some *target* or *response* variable. The set of practical applications is innumerable. Here are some examples:

- Predict how well a user will like a certain documentary on Netflix based on both his viewing history as well as other users' viewing history and how they rated the film on a five-point scale.

- Classify a tumor as a certain type of cancer based on functional magnetic resonance imaging (fMRI) and subject blood markers.

- Predict the price of a stock six months into the future on the basis of company performance measures and economic data.

- Based a customer's shopping cart on a retail website, also suggest additional items that are frequently purchased at the same time to the customer before checking out.

- Predict the likelihood of convict recidivism (when a convict is reincarcerated) based on attributes, including the person's number of previous incarcerations, his age at release, whether the last sentence was for a felony, the type of crime, and whether there are records indicating that the individual has had past problems with alcohol or substance abuse.

These applications are an example of *supervised learning* (Figure 7-7). In each situation, the *training* data set (or sometimes training set) consists of one or more input variables in addition to an observed target variable. Using the metaphor of a student and a teacher, the student would be the algorithm, and the training data would be the teacher when searching for meaningful structure in the data. Also like many students, if presented with additional quality examples, the algorithm will adapt and learn to identify more patterns. As additional observations are collected and additional examples are added to the training set, the algorithm will further adapt and update the way in which it makes predictions.

FIGURE 7-7
Supervised learning patterns

Reinforcement learning (Figure 7-8) is closely related to supervised learning. In reinforcement learning, examples are learned sequentially, one at a time. Consider the scenario in which a mobile phone user resorts to the following procedure to find good reception in a new locale where cellular coverage is poor. The user wanders around, staring at her phone's signal strength indicator while shouting in the phone, "Do you hear me now?" and carefully listening for a reply. This is repeated until she either finds a place with adequate signal or until she finds the new best place under the circumstances. Here, the information the user receives is not directly telling her where to stand to find good reception, nor is each reading telling her in which direction to move next. Rather, after each move, the user might simply evaluate the effectiveness of the current situation; exploration motivates the choice of where to go next.

FIGURE 7-8
Reinforcement learning exploration

In an *unsupervised* machine learning problem (Figure 7-9), the student must learn without the teacher; only the inputs are observed, and there are no measurements on the target variable. The task, then, is to describe how the data is organized or clustered. For example, the construction of association rules is a popular technique for mining commercial databases. In this context, the observations are transactions, like those occurring at a retail checkout. The variables correspond to every item in the business's inventory; each is binary-valued, with "1" indicating that the item was purchased. The goal is to find sets of products that occur with high frequency over the set of transactions; products that are frequently purchased together will have joint values of one. This information is useful for cross-marketing in promotions, demand planning, customer purchase pattern segmentation, and merchandise design.

FIGURE 7-9
Unsupervised learning clusters

What Machine Learning Looks Like

Now that we have an idea of what machine learning is, let's go through a few examples of what machine learning can look like and how it works in the real world.

AUTOMATION

Though it's not a wearable device, a good example of machine learning as it relates to automation is the Nest "learning thermostat." When you first set up your Nest, you set the temperature, and then for a period of time, you continue to manually adjust it to suit your daily schedules and comfort preferences. After a while, the Nest thermostat recognizes your patterns and adjusts itself without intervention. In this case, the observations are the time of day, and the outputs are the temperature that you prefer and whether you're at home. As Figure 7-10 illustrates, if you're home and it's a Tuesday at 4:00 PM, the Nest looks at all of the times you've been home at that time on Tuesdays and knows that you probably want the temperature at 72°F and then sets the temperature for you.

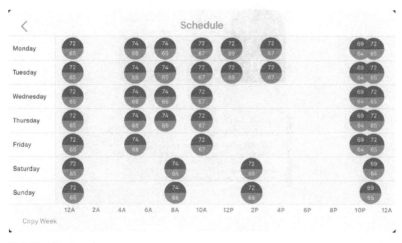

FIGURE 7-10
Nest's schedule

BEHAVIORAL FEEDBACK

The best example of machine learning with wearable devices is Jawbone UP's Smart Coach functionality. As I mentioned in Chapter ?, Smart Coach uses the "Track, Understand, and Act" model, where the Track part is measuring your activity, the Understand part is building a model of that activity, and the Act part is the behavioral feedback. Figure 7-11 presents an example of the type of feedback you might get from Smart Coach.

Let's break down what likely happened to produce this message. After 8 hours and 7 minutes of sleep, the software noticed that the observation of "sleep" was 38 minutes less than the average of 8 hours and 45 minutes. I'm assuming that there's a threshold for when to send a message like this based on deviation from the average, let's just call it 30 minutes. Thus, the software observed that the amount of sleep was outside of the average by +/–30 minutes and consulted the model to pull in predictive information about the outcome of that action, which is the change in the number of steps compared to the average number of steps. The model observed the trend that less sleep means fewer steps, and the observation variable of –38 minutes corresponded to the predicted number of steps of 8,189, which is 542 fewer than the average of 8,731.

FIGURE 7-11
Example of Jawbone's Smart Coach

SHAPING

So how do I, as a Jawbone user, benefit from this message? It's in the recommendation to "Head out and maintain your 8,731 step average." This is an example of *shaping*, a term coined in 1953 by the psychologist B.F. Skinner in his book *Science and Human Behavior*. Shaping is when you identify a target behavior, in this case it's my goal of 10,000 steps, and repeatedly nudge (also known as *differential reinforcement*) the user in small, incremental steps (also known as *successive approximations*) to reach that goal over a period of time. For wearable devices like fitness trackers, behavioral shaping powered by machine learning is the holy grail that we're very quickly approaching.

Here's a hypothetical example of what we could do with machine learning and behavioral shaping when we have a lot more data. Suppose that we have a very popular wearable device with a total of 10 million active users, and the device is able to measure our steps, sleep, caloric intake, and body fat percentage. With all of this information, we can begin performing some supervised machine learning for which our observation variables are behaviors, such as activity, caloric intake, and sleep, and our target outcomes are body fat percentage. The first thing to do would be to group everyone by gender and current body fat percentage, each cohort representing a 2% change in body fat. After everyone was organized in their cohorts, we could analyze the behavior patterns to create a behavior model for each cohort.

The differential reinforcement part of the application would then come from the deviation from your cohort's behavior pattern. For example, we use data from the American Council on Exercise (ACE) to define our target body fat percentages as their "fitness" range, which is 21–24% for women, and 14–17% for men. Deviation toward the behavior model of the lower body fat percentage neighboring cohort would be praised, whereas deviation toward the behavior model of the higher body fat percentage neighboring cohort would be criticized. Each cohort represents a successive approximation toward your overall goal cohort, which is the group with the body fat percentage within the ACE "fitness" percentage range.

TRANSPARENCY

Did anyone else get an icky feeling reading that last paragraph? I think a lot of it comes from the fact that we're not being transparent about the goals or the shaping. It's wrong and paternalistic to shape people's behavior toward a standard about which they have no say. It would be a lot better if, during the onboarding process, we show people their current fat percentage and how that stacks up to the ACE's ranges, and then let each individual choose what body fat percentage is appropriate. Then, during the behavioral shaping process, remind them that the prescriptive feedback that they're getting is only to serve their self-defined goals.

Pulling It All Together

Machine learning is important to service design because this is how we create true services with these devices. Keep in mind that the devices themselves are cheap commodities. The pedometer that my mom wore in the 1990s collects the exact same information as the Fitbit that I bought last week, but the Fitbit stores that information and shows me progress over time. My Jawbone UP3 collects the same amount of information as the 1990s pedometer and the Fitbit, but it actually gives me contextual and personalized feedback that I can use to reach my goals, and that's why I'm wearing it right now, because I actually get something out of it.

[8]

Embodiment and Perception

How We Think

There are competing theories about how we experience our world. The traditional view of cognition is where our understanding of the world is largely metaphorical and the role of our bodies is more or less a vessel for our brains. There's another theory called *embodied cognition*, or *embodiment*, that I find particularly compelling. This theory states that the body plays more of an integrated role in our cognitive process, as opposed to a mere vessel or tool that's controlled by our brains. Embodiment theory claims that the relationship between our mind and our body is more of a two-way street whereby our physicality has more impact on our understanding of the world. This chapter discusses the unique opportunities that wearable devices present in relation to our perception of the world and how to build an understanding of that relationship.

DESCRIBING OUR MINDS

There's a long history of describing both the composition of our bodies and the mechanics of our minds using metaphors based on the prevailing technology of that time. In his book *In Our Own Image*, George Zarkadakis outlines the history of these metaphors in great detail. Zarkadis begins with ancient Jewish and Greek creation myths wherein people were created from mud or clay, and gods would put souls in the clay sculptures to make people. The second paradigm began in the third century BC and influenced the following 16 centuries during which hydraulic pneumatic systems, elaborate systems of tubes that pushed water around to power clocks and other technology, were invented. It immediately began to bleed into contemporary medicine when Hippocrates described the workings of the human body as a ratio of four different types of fluids, or "humors" (Figure 8-1):

blood, yellow bile, black bile, and phlegm.[1] The four humors not only described physiological conditions, but mental conditions, as well. For example, with excess blood, you would be more courageous and hopeful; with excess black bile, you would be despondent and quiet.[2]

FIGURE 8-1

The four humors

With the advent of mechanical, spring-driven devices, these metaphors revolved around describing the mind as a complex, wheel-driven mechanism. The paradigm shifted again in the presence of electricity when Italian physician Luigi Galvani conducted a series of experiments in the late 1700s in which he miraculously reanimated the legs of dead frogs by passing electric current through them. This sparked the use of the phrases "animal electricity" to describe the human body as being driven by electrical current such that the "spirit of life" was electrical in nature. A great example of this thinking is Mary Shelly's *Frankenstein*, in which the monster that protagonist Dr. Frankenstein assembles is given life through the electricity from lightning.

---

1 Leonhardt Thurneysser zum Thurn, *Quinta Essentia* (Leipzig, 1574).

2 George Zarkadakis, *In Our Own Image: Will Artificial Intelligence save or Destroy Us?* (Rider, 2015).

We assemble our current metaphors for the human mind from the parts of a computer and how computers process information (Figure 8-2).[3] We have the hardware—our bodies and brains—and software—the information stored in our brains. We have input via our senses and we collect data. That data then goes to the brain, passing through various processors, and then moves to working-term memory (RAM) where we bring everything together and make decisions. Interpretations are stored in our long-term memory (hard drive) as codified knowledge. Outputs, the actions we take based on decisions, are then generated as a result of this codified knowledge.

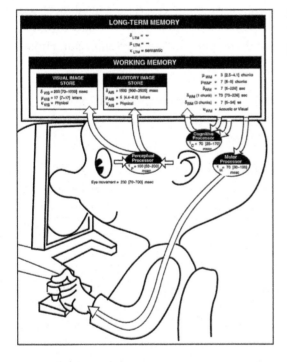

FIGURE 8-2

The model human processor

If we look back over the history of the metaphors that we used to describe our bodies and minds, they seem ridiculous! Of course, we're not governed by ratios of liquid, but these metaphors have real-world repercussions when we decide that the best way to cure a headache is

3   S.K. Card, T. P. Moran, and A. Newell, "The Model Human Processor: An Engineering Model of Human Performance," in *Handbook of Perception and Human Performance. Vol. 2: Cognitive Processes and Performance*, eds. K. R. Boff, L. Kaufman, & J. P. Thomas (Wiley-Interscience, 1986), 1–35.

to cut your arm open to let the extra blood humor out. In another 20 years (or less), we're probably going to collectively move on to a new paradigm and see the computational model of our minds to be just as silly as the pneumatic one, so it's important to be critical of these things when we're applying them to our design process. Thus, the reason why I find applying the approach of embodiment to understanding our minds particularly useful in the design of emerging technologies.

What Is Embodiment?

Most of what I know of embodiment comes from Andrew Hinton's book *Understanding Context*, in which he describes embodiment theory in relation to what he calls *mainstream* or *disembodied cognition*, which I just described as the computer metaphor. Simply put, Hinton states:[4]

> the embodiment argument claims that our brains are not the only thing we use for thought and action; instead, our bodies are an important part of how cognition works.

What this really means is that instead of our thought being confined to our minds, there are a number of other things that contribute to cognition, and those things exist outside of our minds. It means that our environment, sensory information, and our physical actions directly contribute to our ability to process information.

My favorite example of embodiment comes from experience designer Karl Fast's talk at the UX Week conference in 2015, during which he told a story about his son learning about anagrams (rearranging letters of words to create new words with those exact letters) and wanting to create anagram names for everyone in the family.[5] Fast described multiple ways to figure out anagram names: for example, you could just think of the letters in your head and try to figure out the different combinations, you could just write out your name on a piece of paper and try to figure it out that way. But Fast's family ended up taking out

---

4 Andrew Hinton, *Understanding Context: Environment, Language, and Information Architecture* (O'Reilly, 2015).

5 Karl Fast, "Understanding Embodiment," Speech, Adaptive Path UX Week, San Francisco, August 28, 2015.

letters from the board game *Scrabble* and physically rearranging them. He describes the mental process of rearranging the letters, where he'd be shifting them around, certain letters usually go together in words, he kept arranging and found that he could spell the word "star," so he kept that and was left with A, K, F, and L. Fast rearranged those until they resembled something that sounded like a word and ended up with the anagram name Star Flak! What happened was an ongoing loop of action and perception; he would arrange the block, perceive them symbolically, and keep moving the symbols around and reassess how he perceived them until he found his solution.

This is the exact same thing we do in the design process during synthesis, when we're up at a whiteboard arranging pieces of information, trying them out in different groups, and reassessing whether they fit together (Figure 8-3). When we're sketching out scenarios, physically externalizing abstract thought gives us a better understanding of that thought: it allows us to see details and reassess our concepts by physically juxtaposing things. What we're really doing here is creating an external feedback loop to contribute to our understanding of a complex process. That feedback loop consists of external, physical representations of information that we physically manipulate and restructure to assess and reassess the relationships between the pieces of information in order to understand the information as a whole. Simply put, we're creating a cognitive tool that exists outside of our minds.

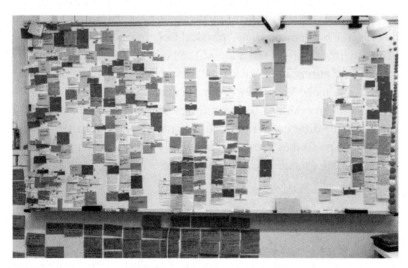

FIGURE 8-3
Whiteboard synthesis during design process

Cognition and Tools

The tools that we use become an extension of our bodies and minds. A small-scale example of this is that I only remember telephone numbers that I knew before cell phones were a thing. I still remember a handful of phone numbers, but they're my family member's land lines, most of which don't exist anymore. Now that phone numbers are stored in phones, you don't need to dial them anymore; you just tap on the name of the person you want to call. This means that the technology that I carry around with me all the time has replaced a function that my brain used to handle, so in a way, the phone in my pocket is now a part of my brain.

A large-scale example of our current cyborg status is what's known as *the Google effect*, or *digital amnesia*. The Google effect comes from a paper published in 2011 by Betsy Sparrow called "Google Effects on Memory: Cognitive Consequences of Having Information at Our Fingertips."[6] The paper outlines a series of studies that looked at how well people recall different types of information. Sparrow concluded that people tend to forget things that they know they can easily search for and find on the internet. Psychologists call this *transitive memory*. This is when we remember where or how to find certain pieces of information instead of retaining that information ourselves.

Not only do tools extend our knowledge of the external world, but they also extend our (understanding of our) bodies themselves. In 2009, the French research group Inserm published a paper showing that, quite literally, the brain incorporates physical tools into what's known as the body schema, the representation of the body in the mind.[7] The study consisted of observing participants' movement when grabbing objects without any tools. Then they had participants use a mechanical grabbing tool for an extended period of time and then removed the tool and measured the differences in the same nontool movement after the tool was used. The studies showed a statistically consistent change in body movement after the tool was removed.

---

6 B. Sparrow, J. Liu, and D. M. Wegner, "Google Effects on Memory: Cognitive Consequences of Having Information at Our Fingertips," *Science* 333, no. 6043 (2011): 776-778, doi:10.1126/science.1207745.

7 Lucilla Cardinali, Claudio Brozzoli, and Alessandro Farnè, "Peripersonal Space and Body Schema: Two Labels for the Same Concept?" *Brain Topography* 21, no. 3-4 (2009): 252-260, doi:10.1007/s10548-009-0092-7.

You could look at this in one of two ways. There's the more conservative view by which you might say, "That's awful! We're reliant on technology, and we used to do everything on our own!" But I prefer to look at it as, "I used to be able to remember 10 phone numbers tops, but now I can remember thousands of them!" or "My hand doesn't hurt from trying to drive this nail in to the wall with my hand!" I'm going to continue this chapter assuming that you share my enthusiasm in relinquishing large parts of our cognitive abilities to the electric overlords, and discuss ways that we benefit from embodied technology in the form of wearable devices.

Embodied Wearable Devices

This understanding of cognition is important to the design of wearable devices because these devices are uniquely positioned to influence the feedback loops that contribute to our mental processes. How this plays out is different for different types of devices; each device has access to specific types of information and different abilities in terms of communicating that information. How you would go about developing a device with these specific inputs and outputs really depends on the kind of service that you want to create. The sections that follow present a few examples of current devices that use sensor-based inputs to give context to situations.

SPIRE

We talked about Spire (Figure 8-4) in Chapter 6. It's a device that monitors your mental state by interpreting movement in your chest or stomach that's caused by breathing. The cycles of information begin with detecting the movement itself, interpreting specific patterns of movement as either calm, focused, or tense, and then reporting that information back to the user. In terms of service that the device provides, there is specific value to the outputs in different time scales. If the device identifies a period of focus, there's no reason to immediately alert the user that she's focused, because, obviously, that would interrupt her focus. If the user is calm, there's no reason to tell her right away that she's calm. The outputs for those conditions *are* valuable, but not at the time of the action, so they're not reported to the user in real time but are able to be observed after the fact in the application.

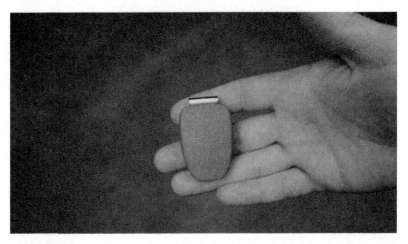

FIGURE 8-4
Spire

Where the Spire contributes to the contextual feedback cycle in real time is when it detects a period of tension. The primary service that Spire provides is giving users an immediate notification of when they're tense. This makes it possible for the users to recognize the things in their lives that cause tension and address the physical symptoms of that tension. Because this part of the service is in real time, the outputs for this case are different: specifically, the Spire device vibrates to signal that you're tense.

LUMO LIFT

The Lumo Lift is a very simple device that you clip to your clothes that alerts you when you're slouching. To configure it, you clip it on your shirt, open the application on your phone, stand up perfectly straight and set that position as the correct posture. When you slouch, it vibrates. The service observes behavior that's usually unconscious (you're not actively making the decision to slouch) and makes you consciously aware of it in order to create the desired behavior of standing up straight.

WITHINGS ACTIVITÉ

The Withings Activité (Figure 8-5) collects the same amount of information as a basic fitness tracker, it looks at steps and sleep cycles. Generally, activity tracker feedback loops are not very immediate. Most people look at their steps after a certain number of days or when their phone notifies them that they only have a certain number of steps to meet their goal. In terms of awareness of progress during the day, if the fitness tracker has a display on the device itself, most of the time the display is off, and you must press a button to turn the screen on to check your progress actively.

FIGURE 8-5
Withings Activité Steel

The Activité differentiates itself from most activity trackers not in its input, but in its output. The physical display of the device has some distinct advantages in relation to embodiment because the "display" is always on. Regardless of whether you're actively checking your steps, that information is there and more directly available to you. Aside from being always on, the "state" of the display doesn't change. It's the exact same information that's available to you in the exact same place.

SMARTWATCHES

In its current form, the Apple Watch doesn't really give us very much contextual information about our immediate environment; rather, it supplies information about our digital lives. Sure, you get the "You should stand up now" message, but that's about it. Very rarely do you get a piece of timely information from your smartwatch that directly informs your perception and action. There are a few exceptions though; for example, Google's Androidwear is one of the few popular smartwatches that take advantage of geolocation-based information with which you can set a reminder to get milk when you're physically near a grocery store.

Something very different in terms of immediate contextual information is a watch specifically designed for outdoor sports: the Suunto Ambit3. The Ambit3 is a GPS-enabled watch that is built around giving you very specific contextual information when running, hiking, skiing, or doing any other outdoor sport. It keeps track of your distance, pace, speed, and exact location on preplanned routes, giving you immediate contextual information about your environment and performance. In addition to status-type data, the Ambit3 also has certain messaging features such as warnings when your heart rate exceeds what it should be, telling you to take a break, as well as monitoring changes in barometric pressure, warning of an approaching storm.

POKEMON GO PLUS

The Pokemon Go Plus is an interesting new addition to the wearables landscape. It's designed to vibrate whenever you're near a virtual Pokemon monster that you can collect in the game Pokemon Go, in which you collect virtual monsters that are distributed in the physical world. Oddly enough, even though it's a game, this is a great example of a wearable device that gives you actionable contextual information about your environment, relying on both vibrations and flashing light signals to inform users of an element of their environment that they would otherwise be unable to perceive without their smartphone.

Perception

With wearable devices, we're working in the "real world," and that means that there are a lot more variables available to us because we're not limited to the screen. Not only are we designing in an environment that has an extra dimension, that extra dimension can be perceived in many different ways! We're dealing with an exchange of information through a variety of senses simultaneously, so this becomes our new platform, and I think it's a good idea to look into how that platform—the body—understands the world around it.

EMBODIED PERCEPTION

In terms of perception, one of the main differences between embodiment and mainstream cognitive theory is the order in which things happen and how we think about the structure of comprehension. The mainstream view is anchored by the idea that our understanding of the world is reliant on the symbolic abstraction of concepts, wherein we receive information through our senses, process them abstractly in our minds, and then take action based on those abstractions. In embodiment, the symbolic comprehension happens only after the experience. Andrew Hinton gives the example of a bird flying:

> A bird doesn't "know" it's flying in the air; it just moves its body to get from one place to another through its native medium. For we as humans, this can be confusing, because by the time we are just past infancy, we develop a dependence on language and abstraction for talking and thinking about how we perceive the world—a lens that adds a lot of conceptual information to our experience. But the perception and cognition underlying that higher-order comprehension is just about bodies and structure, not concepts. Conscious reflection on our experience happens after the perception, not before.

In the book *Self Comes to Mind*, neuroscientist Antonio Damasio outlines his theory of how our body-first cognition came about. He looks at the behavior of simple single-cell organisms which have "what appeared to be a decisive, unshakable determination to stay alive for as long as the genes inside their microscopic nucleus commanded them to do so."[8] He continues to explain that as we evolved in to multicellular organisms, this pattern continued, and even in our current state, most of the decisions we make are to maintain *homeostasis*, the narrow range of physical conditions required to sustain life. He proposes that even higher-level thought arose from our awareness of these decisions, or, as he puts it, making "the basic life-management know-how, well, *knowable*."[8] Whereas birds and other animals might not know exactly why they're doing things, we are aware of the reasons, but that knowledge precedes our conscious experience of that knowledge.

---

8 Antonio R. Damasio, *Self Comes to Mind: Constructing the Conscious Brain* (Pantheon Books, 2010).

An interesting demonstration of sensory information bypassing our conscious experience is called *blindsight*, wherein people or animals suffer partial damage to their visual cortex and are no longer able to experience sight, but still react to visual stimulus. This was first observed in macaque monkeys, who were once able to see, lost their ability to consciously perceive sight, but were able to react to visual target stimulus on a screen.[9] These types of studies have been repeated many times in humans. Most famously, a study by Harvard neuroscientist Beatrice de Gelder demonstrated that a man who had gone blind as a result of two strokes was able to walk without assistance down a hallway full of obstacles and consistently avoid every one of them.[10]

Breaking Down Perception

When we think about sensory perception, it seems pretty simple: we see things, we hear things, we touch things. But it's not that simple if we try to get detailed about what's really going on between our minds and the outside world. I tried to answer this question by asking a couple more questions: What are we experiencing? How much of what we experience is being actively perceived? How much of what we're perceiving is objective reality? The answers to these questions can't really be quantified (yet), and when asked, my cognitive science PhD buddies would tell me what they thought, but wouldn't let me directly quote them on the subject because they don't want to make public claims.

On a really high level, here is how I understand perception: we get separate "images" from each of our senses, all of them at the same time. We get those images somewhere between 60 to 100 times per second (60–100 Hz), we have the ability to actively perceive maybe 3% of that, and of that 3% we have the ability to actively comprehend maybe 1%. To demonstrate how I arrived at this conclusion, I'm going to focus mostly on our sense of vision, because that's the one that takes up the most processing power in our brain.

9 Bernard J. Baars, William P. Banks, and James B. Newman, *Essential Sources in the Scientific Study of Consciousness* (MIT Press, 2003).

10 Alessia Celeghin, Beatrice de Gelder, and Marco Tamietto, "From affective blindsight to emotional consciousness." in *Consciousness and Cognition* 36 (2015): 414–425, doi:10.1016/j.concog.2015.05.007.

RAW INFORMATION

According to Antonio Damasio, we receive perceptual *images* in the form of two-dimensional grids or "maps" similar to a pixelated image on a computer screen. These grids are mapped directly to the visual inputs in our eyes, the rods and the cones, that act like photo receptors. Humans have about 120 million rod cells that give us raw information about the intensity of light and about 6 million cone cells that give us information about color.

SPEED

In terms of frequency, the cognitive psychologists with whom I've spoken tell me that we receive information about 60 to 100 times per second. This information comes from different attempts to measure how quickly our neurons fire, called *neural coding*.

BIOLOGICAL ATTENTION

The first filter on all of that information is what our eyes are physically focusing on, or our *foveal resolution*. The fovea is the part of our eye that has the ability to maintain sharp focus. Foveal resolution is the amount of space to which we have the ability to biologically pay attention. For example, if you have perfect 20/20 vision, you don't have 20/20 vision in your periphery; you have 20/20 vision only on what you're focal point. The amount of space that you have the ability to physically focus on is about the size of both of your thumbnails next to each other if you hold them out at arm's length. Our field of vision is about 95°, and the foveal field of vision is roughly 3°, so this cuts our active perception down to about a little over 3%.

COGNITIVE ATTENTION

The next filter is our ability to cognitively understand the information to which we're paying biological attention. As it turns out, there's a lot of other things going on in your consciousness that are competing for your processing power. For one, even though vision takes up most of your brain, you have four other senses that are also sending you information, not to mention anything else you might be thinking about that doesn't have anything to do with what you're perceiving.

MEMORY AND EXPECTATIONS

In addition to your senses, a lot of your perception is colored by your previous experiences. As Andrew Hinton puts it:

> Our brains evolved to support our bodies, not the other way around. What else would memory have mainly evolved for other than recalling just enough information about our surroundings to help us survive?

I see this all the time just walking around my neighborhood—we really "see" only what we expect to see. For example, a few years ago my friend bought the ugliest car I've ever seen. It was an orange-brown colored subcompact, and I remember telling her that I've never seen that car before. Right after that I began seeing them *everywhere*. Multiple times a day I'd see one and think it was her driving around because she had the only one. But as it turns out, they were all over the place; I just never actually acknowledged them. An easier example of our brain filtering out unnecessary information is that you can see your nose right now. It's always been there. Even if you're looking straight ahead, it's in your field of vision, but you never actually "see" it.

Sensory Substitution, Addition, and Augmentation

There's a lot of information in our environment that we just don't have access to. We can see about 0.0035% of the electromagnetic spectrum, and we can hear very little of the sound that's actually around us. Sensory substitution, addition, and augmentation give us access to different types of sensory experiences that would otherwise take millennia to develop by natural evolutionary means. Though these technologies are still in their infancy, there's a lot of interesting work happening right now that can be useful to someone designing wearable experiences. The sections that follow present a few of them.

AUGMENTATION

We're just beginning to scratch the surface of consumer sensory-augmentation. Certain elements of this technology have been around since the 1200s in the form of eyeglasses, and we've had electronic hearing aids since 1898, but the primary functionality of these technologies have changed very little from their inception. Eyeglasses exist to correct our vision, and hearing aids still amplify sound. More advanced augmentation allows us to fully control, manipulate, and increase the information we're getting from our various senses.

The emerging class of consumer sensory augmentation wearables is led by Doppler Labs and its Here earbuds. These are simply in-ear headphones that have a microphone on the outside and a Bluetooth-enabled microprocessor between the microphone and the speaker. Here goes far beyond simple amplification by allowing you to fully control its audio reality, giving you the ability to turn up or down certain aspects of the sound in your environment. This is really interesting to me because a lot of what we experience is informed by sound. It's kind of a gut check that everything is normal, and it's always on.

Here is different than any other wearable device I've encountered. Doppler Labs describes using its product as "active listening," where sound becomes something with which you actively engage. After you spend a little time clicking through the different settings and playing with the slides, you really understand a lot of what's going on around you. You begin to recognize the low rumble of your building's infrastructure when you completely tune it out, you isolate certain parts of your audio environment that you want to focus on, and then someone taps you on the shoulder because they've been trying to talk to you and you had no idea they were there!

The most interesting part about wearing Here around for a while was how quickly you begin to adjust to this alternate reality. If you're not distorting the sound with something obvious like echoes or flanges, what you're hearing isn't actually that noticeably different until you see something happening and the sound doesn't quite match up as you'd expect. This raises some interesting questions about what sensory augmentation will evolve into, such as what levels of filtration of our environment are acceptable? And how will interpersonal relationships change if we're experiencing the same thing in very different ways? I suppose we'll find out soon.

SUBSTITUTION AND ADDITION

All of our senses are a result of evolution, but what if we could speed up this process with wearable devices? What if we could use devices to give ourselves the ability to physically perceive things that we wouldn't be able to otherwise? This is exactly what neuroscientist David Eagleman is doing.[11] In 2015 he published a study in which he gave deaf people a vest that incorporated a grid of haptic motors in the back (Figure 8-6) that converted sound waves to vibration patterns across the person's back. After about three months of training, the deaf person would gain a direct perceptual experience of hearing, just not through the ears but through the skin.[12]

FIGURE 8-6
Subject writing words that he hears from his back

This is incredibly important for the future of wearable devices for one main reason: Eagleman showed that our brains have the ability to be completely remapped to completely understand new inputs without having to consciously translate them. This means that, for example, we could take detailed information such as telemetry information of an aircraft and feed that information directly to the pilot as opposed to displaying it on a very complicated dashboard, or allow diabetic people to feel their blood sugar levels.

---

11 David Eagleman, "Can We Create New Senses for Humans?" Speech, TED, March 2015 (*https://www.youtube.com/watch?v=4c1lqFXHvqI*).

12 Scott D. Novich and David M. Eagleman, "Using Space and Time to Encode Vibrotactile Information: Toward an Estimate of the Skin's Achievable Throughput," *Experimental Brain Research* 233, no. 10 (2015): 2777–2788, doi:10.1007/s00221-015-4346-1.

Gaining Physical Awareness

In 2012, I began working on a series of workshops with my colleague Erik Dahl to explore the relationship between our physicality and our minds. The workshops centered on exploring Laban Movement Analysis, a framework to describe human movement that was developed by Rudolf Laban, an early 20th-century dancer and theorist who is often referred to as "the father of modern dance." More specifically, the workshops focused on the "Effort" section of the framework, which consists of three binary switches that he called analyses: space, weight, and time. Different combinations of these three binaries produce eight archetypal "efforts."

I first came across the Laban frameworks in a previous life when I was studying theater and dance. I took a lot different movement and modern dance classes for years, but 10 years later, the Laban work stayed with me. The classes, which you can see in Figure 8-7, were meant to give us, as actors and dancers, a framework to describe movement, to increase specificity in our decisions on stage. But one of the biggest things that I learned in the class is that the relationship between the body and our emotions and thoughts is a two-way street. After starting these classes, it became abundantly clear that when you feel a certain way, it changes the way you move through space. But it's also the other way around: when you change the way you move through space, it changes the way you feel.

FIGURE 8-7
Laban workshop, 2012

The first time Erik and I participated in a Laban workshop, we were very focused on the emotional aspects of the relationship between our minds and bodies. That made a lot of sense to us initially, but we kept getting very consistent feedback from the workshop participants about how after the workshop they fundamentally understood their movement differently. They'd say things like "I found myself being a lot more mindful of my actions on my way home. I thought about how I interacted with my car more. From shifting, to how I felt in the seat. I was just more aware of things—things that I just generally do on autopilot. But the workshop brought me back to a state of bodily awareness that I don't experience very often." Another person commented, "As I walked down the sidewalk, it was as though there were a new UI inside my brain–like *virtual reality*,' except it was just 'reality'–and the meters in the heads-up display had labels of weight, space, and time."

After developing the workshops a bit more, we found that the framework was really useful for gaining a certain sensitivity for interactions that happened in physical space. When working on Internet of Things (IoT) and wearable projects that happened in physical space, we were able to recognize and describe things that we didn't really notice previously, allowing us to understand and correct our design work in a more concentrated and intentional way.

FRAMEWORK STRUCTURE

The Laban framework has two levels. There are *analyses*, which are binary concepts that describe human movement through space in a more mechanical way. Those binary choices are combined to form *efforts*, which describe more complex movement.

Analyses

The framework begins by breaking down physical movement in to three dimensions, or "analyses." These analyses are *weight, space,* and *time* (see Figure 8-8). Weight is broken down into *light* and *strong,* which describe the physical force of the movement; for example, a ballerina on her toes might be lighter, and a soldier marching might be on the strong end of the spectrum. The space analysis refers to the path of your movement and is described on a spectrum from direct to indirect. This refers to both the movement of the individual parts of your body as well as the path that your body takes through the space. The third analysis is time, which refers to your pace on a scale of sudden to sustained; sudden is generally faster or more erratic movement, and sustained is more consistent and usually slower movement.

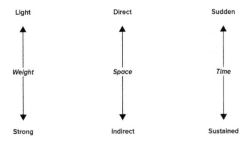

FIGURE 8-8
The three binaries of the Laban Effort Framework: weight, space, and time

The workshop commences by exploring these binaries individually. We begin walking around as we normally would, take note of how that feels, and start paying attention to how our feet are hitting the ground, how our arms are moving, and our pace. We then cycle through light, strong, direct, indirect, sudden, and sustained individually.

Efforts

The efforts come together through combining different ends of the analyses. After cycling through the individual analyses, we begin putting them together. For example, we go back to moving lightly. After we've figured out what that means to us, we move lightly and directly. After a few moments of experiencing what that feels like, we add the third analysis and move light, direct, and sustained. Putting the three analyses together creates an effort, and moving light, direct, and sustained is called *Gliding*, as illustrated in Figure 8-9.

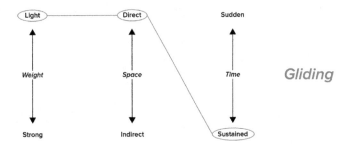

FIGURE 8-9
Light, direct, and sustained movement is Gliding

Next, we begin switching the binaries. We go from light, direct, and sustained to *strong*, direct and sustained, this brings us from Gliding to *Pressing*, as shown in Figure 8-10.

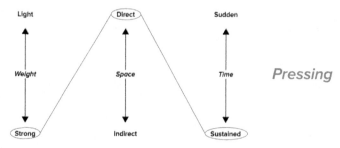

FIGURE 8-10

Strong, direct, and sustained movement is Pressing

We continue to cycle through all eight of the efforts (Figure 8-11, Figure 8-12). With each one, we take a few minutes to think about how we're moving and think about our feelings and how that effort changes our emotional state and our mental state. For each effort, we explore the movement with different constraints. We might begin with Pressing in the most extreme and exaggerated form possible and then try to make it even bigger and more exaggerated. Then, we bring it down a little bit. We imagine we're in our office building or in a public place where it would be weird to do exaggerated movements, and we'll press in a normal place. We'll isolate certain parts of our bodies: what does it mean to Press with only our eyes?

Weight	Space	Time	Effort
Strong	Direct	Sudden	Punch
Strong	Indirect	Sudden	Slash
Strong	Direct	Sustained	Push
Strong	Indirect	Sustained	Wring
Light	Direct	Sudden	Dab
Light	Indirect	Sudden	Flick
Light	Direct	Sustained	Glide
Light	Indirect	Sustained	Float

FIGURE 8-11

The eight efforts

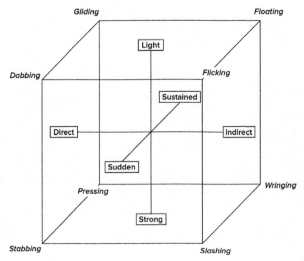

FIGURE 8-12
The eight efforts expressed dimensionally, known as the Dynamosphere

Scenarios

After exploring the framework and getting comfortable with the efforts for the first half, the second half is dedicated to embodying the efforts in different scenarios. This part of our workshop has evolved a lot over the years. Early in the evolution of the workshop we would hand out little pieces of paper with an effort on it to everyone in the workshop and have them move around in the space. After everyone is in the space, we ask them to take note of their effort and how interacting with other efforts would influence their movement or feelings. We'd take the scenarios further by taping off large sections of the space and confining people to increasingly smaller spaces to encourage more direct interaction.

The more recent iterations (Figure 8-13) are rooted in more specific scenarios that more closely represent the real world. One of the scenario-based exercises involves the interaction between two people who are asked to complete a joint spatial task while beginning the task with different efforts. This is more directly applicable to the aspects of service design that exist in dimensional space such as devices or human-to-human interaction. Other exercises involve observing the differences in physical behavior of a person using a device or in a specific scenario and then changing that device or environment to create a more appropriate physicality.

FIGURE 8-13
Laban workshop at Adaptive Path (2015 photo by Iran Narges)

Pulling It All Together

There are a lot of things to be aware of when designing devices and services for the physical world, and it is extremely important to be aware of how our brains perceive the world around us when we introduce new stimuli. Having a basic understanding of cognition, perception, and physicality is an asset in conventional wearable design, but also opens up a new realm of possibilities moving forward, extending our natural senses and allowing us to perceive the world around us in new ways.

Prototyping

Prototyping is a big part of the design process for anything. I come from the world of experience design, and I couldn't imagine a project for which we don't figure out if we're building the right thing and eventually work out the kinks of a product by using prototypes. Prototyping lets you quickly build a lower-fidelity (not fully functional but enough to learn from) version of your project to get a better understanding of what you're building, and eventually get feedback from people in terms of how they understand it. This chapter covers the why part of prototyping, the tools, and processes that are available to prototype wearable devices. We even dive in to some of the technical aspects involved.

Why Prototype?

I feel like a lot of people see the role of prototypes as something that you use for feedback on a design, mainly in user testing. In fields of design with a more established language, you can get by with using prototypes only in this way. I could verbally describe an application to someone, and because the design elements that we use for two-dimensional design are so established, you could probably get a good feel for the idea this way. In this newer world of wearable devices, though, it's a little different. Not only is there no established common language, but the products themselves are so divergent and new that we can't even know that it's a good idea to begin with without prototyping it to make it real in some way.

In the world of traditional two-dimensional design, you can draw a picture of something and convey a lot of what's going on with the product. In fact, a two-dimensional image of the design has a lot in common with the product you're describing because of the physical limitations of the platform for which you're designing. In designing wearable devices, by the nature of what you're building, those limitations don't exist, because you're designing a service that only exists over time and isn't contained to the device or devices that you're using to enable the service. In most cases, the devices that you're building to support the service are only a small part of the whole, and they exist more as a symbol of the service in the minds of the users.

WHAT DOES THAT LOOK LIKE?

There's no single answer to the question, "What does that look like?" because the types of wearable devices we've covered in this book are so different. But we can break down the big parts of the prototyping process that are going to be common to a lot of wearable devices, such as sensors, connectivity, and data. What we're going to be doing here is chaining together a broad set of skills and technologies to take a service from an abstract idea to something tangible and from which we can learn. To do this, we're going to have to get pretty technical in terms of programming and electrical engineering, which can be intimidating, but you can do it!

So, here's the part where I say the words, "You need to learn to code," and we both let out a sigh because we've collectively been having this conversation forever and we're both tired of talking about it, so I'll try to be brief. You need to learn to code to prototype these products; however, the type of programming that you need to learn is not the same as a traditional developer. The type of code that I'm talking about here is a sketching medium that comes from a place of exploration and will not directly translate to the production code that will eventually power what you're building. These skills are great for all kinds of design, and on top of everything, it's fun!

Foundational Tools

The two primary tools that I use for this type of prototyping are Processing and Arduino. Processing is an open source programming language. Its development environment is based on Java and was originally built for designers and artists. Arduino is an open source hardware platform for designers and artists with a development environment based on Processing. I'll get much more specific about the tools later, but the primary reason I believe in these technologies as the foundational tools in your workshop is because of their accessibility, both in terms of community and in learning curve.

I wouldn't say that Processing and Arduino are inherently easy to pick up, but I do believe they're significantly easier than almost anything else because of the incredible community that supports them. There are near infinite resources for all skill levels that are available in nearly any form you'd prefer. And if you have any question or need any help, the forum-based communities are amazingly friendly and helpful. This is what makes these tools so great.

PROCESSING OVERVIEW

Processing is the main tool that I use for sketching. Its main functional advantages are that it's really quick to set up, you don't need to configure much of anything, and that most of what you do in Processing has a visual output so that what you're making is less abstract. When I prototype wearable devices, Processing is the tool that does most of the heavy lifting. For this section, I'm going to go through an extremely brief overview of Processing—this is not meant to teach it by any means; it's just to get an idea of what it looks like.

The Processing IDE

The first thing you need to do is go to *https://processing.org* to download the Processing *integrated development environment*, or IDE (which is where you write the code). After you download it, the Processing IDE unzips as an application. You can install that wherever you install your applications and then open it up. It should look something like Figure 9-1.

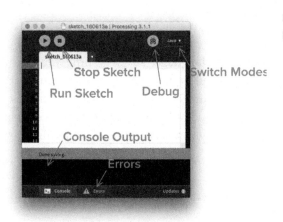

FIGURE 9-1

Processing's IDE

The play button compiles and runs the sketch; the stop button stops the sketch; the debug button opens up a debugging screen; the Java button lets you switch modes (we'll get to the modes part soon); the black part at the bottom is where you see any console outputs; and the second tab is where any errors will show up when you run your sketch.

The structure of a sketch

There are two main parts to a sketch: the *setup* part and the *draw loop*. Between the curly brackets following void setup(), you define some characteristics of your sketch, such as how big the output window will be and the background color of the sketch. In the curly brackets following void draw() is where you put the majority of the instructions for the sketch. This is called the draw loop, because Processing executes the instructions (code) in the draw loop from beginning to end, and then it loops back around to the beginning and does the same thing. This happens about 60 times per second. In Figure 9-2, the setup part runs once, and then the draw part runs over and over again.

FIGURE 9-2

Structure of basic sketch

The sketch output window

Processing is a visual-oriented language, so every time you run a sketch in it, there's an output window that pops up to show you what's going on, as demonstrated in Figure 9-3. In the preceding example, we defined the size of this output window in the setup section as 500 pixels wide by 400 pixels in height, with a background of white, and drew a line in the draw loop.

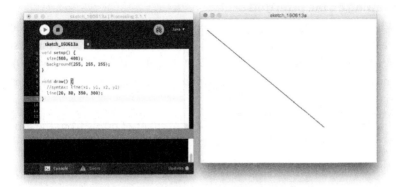

FIGURE 9-3
Screenshot of basic sketch with output window

What the code looks like

Figure 9-4 shows that for a lot of the code instructions you'll be writing, there are two main parts: the function (line, in this case), and the arguments (the numbers), which are specifically structured pieces of information that you type in the parentheses.

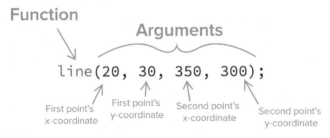

FIGURE 9-4
Diagram of line instructions

Figure 9-5 presents a diagram for line, which draws a line between two points. The four arguments are the x-coordinate for the first point; the y-coordinate for the first point; and then the x-coordinate for the second point; and, finally, y-coordinate for the second point. In the example, those values are 20, 30, 350, and 300.

Drawing on the pixel grid

When drawing things to the screen, everything (two-dimensional at least) is drawn on an x–y pixel grid. As Figure 9-5 illustrates, the first pixel in the upper-left corner is (0,0), the one to the right of it is (1,0), and the one below it is (0,1). When you're drawing something, it helps to think of its position as follows:

("number of pixels from the left side," "number of pixels from the top)

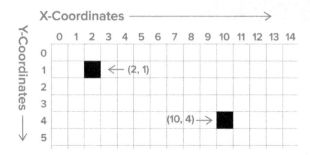

FIGURE 9-5
Diagram of pixel grid

In the preceding example, we drew a line that begins at a grid position 20 pixels from the left side and 30 pixels from the top, and it ends the line at 350 pixels from the left side and 300 pixels from the top (Figure 9-6).

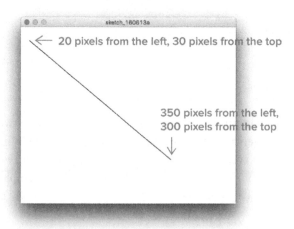

FIGURE 9-6
Points of the line with pixel locations labeled in the output window

Layers

Every time the draw loop is run, it draws what you define on top of the previous layer (Figure 9-7). In the example, the first instruction of the draw loop is a background color that gives the impression of a clean slate every time. If that weren't there, you'd see everything the sketch has ever drawn to the screen.

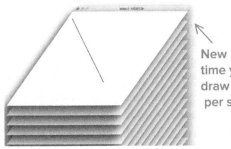

New layer is drawn every time you cycle through the draw loop (about 60 times per second)

FIGURE 9-7
Diagram of layers

Reference and examples

All of the functions you can use in Processing are listed in the reference page on the Processing website. It's incredibly useful and shows you how to use each function with a short chunk of example code. You can also highlight a function in your sketch, right-click the highlighted function, and select "Find In Reference," which will take you to that function's page in the reference. There are also a large number of examples. To access them, on the menu bar, click file, and then click examples. These are really helpful when you're trying to do something and you need to see how particular commands and functions work.

Libraries

Processing Libraries are really useful ways to extend the functionality or make some advanced tasks a little easier. To add a library to your sketch, on the menu bar, click the Sketch drop-down menu, and then select Import Library. There are a handful of libraries that ship with Processing. Also on the menu bar, you can click Add Library, which lets you add all kinds of great third-party libraries. I suggest starting out with Andreas Schlegel's controlP5—a great graphical user interface library for Processing.

Android mode

Android mode is one of my favorite parts of Processing. You can use it to export your processing sketch to an Android device. To switch to this mode in the upper-right corner of the IDE, click Java, click Add Mode, and then select Android mode. Then, plug in an Android device to your computer, and when you run the sketch, it will run on an Android device! I'll get a little more in depth with this toward the end the chapter.

Getting comfortable with processing

There are seemingly endless resources out there to help you learn Processing. Of course, everyone learns in different ways, but there are a couple resources that really helped me become comfortable with Processing. First is just going through the examples that ship with it. They're small, complete sketches that explain a lot of what Processing can do. I'd start by just clicking through a few of those to get an idea of what's going on. The second resource is anything that comes from Daniel Shiffman, an amazing creative coder and professor in NYU's ITP program. Daniel's "Orange Book," *Learning Processing*, is the primary way I was able to learn Processing so quickly, and I still reference it from time to time when I need to do something that I haven't done in a while. Additionally, Daniel has an incredibly comprehensive video series hosted on his Vimeo account at *https://vimeo.com/shiffman*.

ARDUINO OVERVIEW

On a high level, an Arduino board (Figure 9-8) is a very simple computer that takes an electrical input (usually from a sensor, but it also could be from a device, computer, or transmitter), does some form of computation, and outputs a corresponding electrical signal to some device (a motor, a transmitter, etc.). For example, I could configure a motion sensor to trigger a light to come on by setting up the Arduino so that whenever it received an input from the motion sensor, it would send power to the light. I could also turn on the light by hooking my Arduino up to a Bluetooth antenna that I could trigger from my phone.

The Arduino board

At the top of the Arduino Uno that you can see in Figure 9-9, there's a place to plug it in to a power outlet, and the serial port with which you plug it in to your computer. The rest of the Arduino board comprises a series of contact points called pins. In the Power and Ground section above the pins are labeled GND, 3.3 V, and 5 V. You use these pins to provide power for a sensor that you plug in to the board. Below the Power and Ground pins are the analog pins, which are labeled A0 through A5. This is where you can plug in analog sensors. These pins are input only; the board doesn't have analog output. On the opposite side are the digital pins, they're labeled 0 through 13 and can function as both input and output ports. You can use the digital pins with the tilde (~) next to them for what's called *pulse width modulation*, which is when you mimic an analog output by turning the digital pin on and off so quickly that it seems like an analog output (e.g., it turns an LED on and off so quickly that it looks like it's dimming).

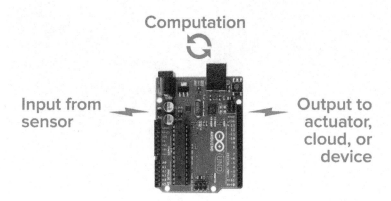

FIGURE 9-8
The Arduino Uno, the most common Arduino board

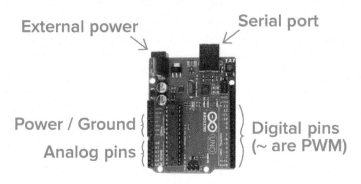

FIGURE 9-9
Inputs and outputs of the Arduino board

Different boards

I'm using the Arduino Uno in my example because it's the most common Arduino board. But, one of the wonderful things about open source hardware is that you can make your own boards! This means that they come in all shapes and sizes, as depicted in Figure 9-10, with different built-in capabilities. I'll go through a few of these toward the end of the chapter.

FIGURE 9-10
A variety of different Arduino boards

The Arduino IDE

The Arduino IDE (Figure 9-11) is very similar to that of Processing. There are the buttons at the top that start the code, and there is a console output on the bottom. The big differences are that Arduino has more buttons for starting a new sketch, opening an existing sketch, and saving the current sketch. You'll also notice that there's no button to stop the code. This is because after the code is loaded on to the Arduino board, it will keep running over and over again as long as the board has power—it doesn't need to be still plugged in to the computer.

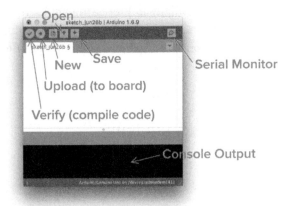

FIGURE 9-11

The Arduino IDE

Arduino code

Similar to the structure of the sketch for Processing, the Arduino sketch has two main parts: the setup part and the loop (Figure 9-12). Between the curly brackets following void setup(), you define some characteristics of your sketch, such as stating what certain pins will be used for. In the curly brackets following void loop() is where you put the majority of the instructions for the sketch. Just like Processing, it executes the instructions (code) in the draw loop from beginning to end, and then loops back around to the beginning and does the same thing over and over again.

FIGURE 9-12

Basic Arduino sketch structure

The code itself is also structured similarly to Processing with a function, arguments in parentheses, and semicolon line endings. In this line shown in Figure 9-13, we're configuring digital pin number 11 to turn on.

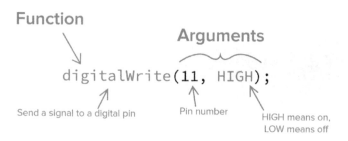

Function

Arguments

digitalWrite(11, HIGH);

Send a signal to a digital pin   Pin number   HIGH means on, LOW means off

FIGURE 9-13
Diagram of line instructions

Sensors

Sensors are obviously a huge part of wearable devices, so before we get more detailed about how Arduinos work, let's talk about what sensors actually do. Generally, sensors take some kind of physical input like touch, motion, light waves, or audio waves, and converts those signals to an electrical signal.

As Figure 9-14 shows, there are two main types of sensor outputs that the Arduino can receive: digital signals and analog signals. Digital signals are either on or off (represented in code by a 1 or a 0, or high signal or low signal), kind of like a light switch.

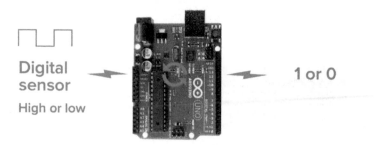

Digital sensor

High or low

1 or 0

FIGURE 9-14
Digital sensor input and output

Analog signals (Figure 9-15) contain a lot more information. Analog sensors that work with the Arduino usually output an electrical signal somewhere between 0 and 5 volts. Then the Arduino converts that input to a digital signal that returns a value between 0 and 1,023. In short, digital sensors are either on or off (two states), and analog sensors have 1,024 different potential states (0 is also a potential state, thus the total of 1,024).

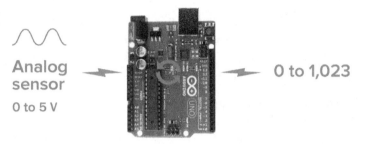

FIGURE 9-15
Analog sensor inputs and outputs

Short example

For an incredibly short (only two lines of code!) self-contained Arduino example, we're going to take an analog sensor and make it dim an LED. Figure 9-16 shows what it's going to look like.

FIGURE 9-16
Everything hooked up

For this example, I'm using my favorite analog sensor, a MaxSonar EZ1 by Maxbotix. The EZ1 is an incredibly simple ultrasonic range finder. It sends out ultrasonic audio waves, listens for them to bounce back off of something, and then converts that distance to a voltage between 0 volts and 5 volts. I plugged the GND pin on the sensor to the GND pin on the Arduino, and the +5 pin on the sensor to the 5 V power to provide power to the sensor. I then plugged the AN pin on the sensor to the A0 pin on the Arduino.

To dim an LED, you need to plug it in to a PWM pin with the tilde (~) symbol next to it on the board. For this example, I took a white LED and plugged the long lead into the 11 pin and the short lead in to the GND pin that's right next to the 13 pin.

The code is incredibly simple:

```
void setup() {
  pinMode(11, OUTPUT);
}

void loop() {
  analogWrite(11, analogRead(A0)/4);
}
```

The first line of code in the setup section declares that we're going to use pin number 11 as an output.

The second line of code in the loop section instructs the Arduino to write an analog signal to pin number 11 by using the analogWrite function. The first argument for this function is 11, which specifies that pin number 11 is the one to which to write. The second argument, analogRead, is also a function that instructs the Arduino to take the signal from the A0 pin which is the pin that we've used for the sensor. The signal is going to come in to the Arduino as a number between 0 and 1,023, so I divided that number by 4 in order to map it to the PWM pin, which takes an input between 0 and 255.

If everything is hooked up properly, you'll be able to move your hand closer and further away from the sensor to make the LED become brighter and dimmer, respectively.

Getting comfortable with Arduino

Similar to Processing, there are many resources out there that you can use to learn all about Arduino. A great start is definitely to poke around with the examples that come with the IDE and maybe pick up a starter kit. I always suggest getting everything from *https://www.adafruit.com* because it has the most well-constructed learning material for everything it sells. Another great resource is the O'Reilly book *Getting Started With Arduino*, which is very helpful when you're first starting out.

PROCESSING AND ARDUINO TOGETHER

I have a confession to make: I very rarely ever use Arduino by itself. For me, it's a lot easier to just pipe the data from the Arduino board directly to Processing so that I can visualize the data in a way that isn't just blinking an LED. I think you'll find this particular setup a lot more useful if you're prototyping a wearable device.

Putting Firmata on your Arduino board

The first step is to load the Firmata sketch on to your Arduino. Firmata is a collection of software libraries for both Processing and Arduino that lets them interface with each other. Firmata ships with the Arduino IDE and you can find it by going to the File menu in the Arduino IDE, selecting Examples, and then, in the "Examples from Libraries" section, select Firmata and click StandardFirmata, as illustrated in Figure 9-17.

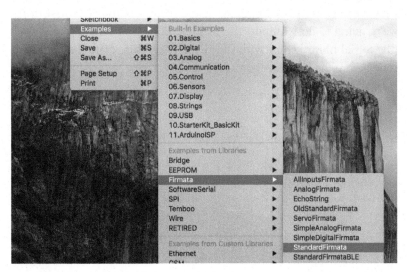

FIGURE 9-17
Finding Firmata

This will bring up the Standard Firmata sketch. Load that on to your Arduino, and then close the Arduino IDE, (if you don't close out of the Arduino IDE, Processing won't be able to communicate with your board).

Connecting to Processing

Now, you need to download the Firmata library for Processing. To do so, go to the Sketch menu in Processing, click Import Library, and then click Add Library (Figure 9-18).

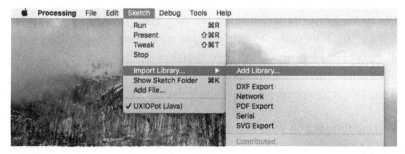

FIGURE 9-18
Firmata in Processing

This opens up the Contribution Manager. Simply click the library with the title of Arduino (Firmata), and you're good to go.

Now we're going to open a new sketch in Processing. The first thing we're going to do is import the Firmata library and the Serial library (the Serial library ships with the IDE). To do this, write the following code before the setup section:

```
import processing.serial.*;
import cc.arduino.*;
```

Next, we need to set up the Arduino and make a variable to store the sensor information:

```
Arduino arduino;
float potRead = 0;
```

Then it's time for the setup section. The code that follows sets the size of the sketch to 300 pixels wide and 300 pixels tall, and the background of the sketch to white:

```
void setup() {
  size(300, 300);
  background(255);
  fill(255, 0, 0);

  for (int i = 0; i <= Arduino.list().length-1; i++)
  println(i + " - " + Arduino.list()[i]);

  arduino = new Arduino(this, Arduino.list()[REPLACE], 57600);

  for (int i = 0; i <= 13; i++)
    arduino.pinMode(i, Arduino.INPUT);
}
```

After you run the sketch for the first time, in the console output window beneath the text editor, you'll see a list of ports that looks something like Figure 9-19.

```
0 - /dev/cu.Bluetooth-Incoming-Port
1 - /dev/cu.Muse-2B90-RN-iAP
2 - /dev/cu.usbmodem1411
3 - /dev/tty.Bluetooth-Incoming-Port
4 - /dev/tty.Muse-2B90-RN-iAP
5 - /dev/tty.usbmodem1411
```

FIGURE 9-19
Console output

You now need to determine which one of these ports are connected to your Arduino. It's going to be one of the ports labeled usbmodem, and you're probably going to need to go back and forth a few times picking different ones. In the preceding code example, look for the line that has REPLACE, delete that word, and replace it with the number in the list in Figure 9-19 that corresponds to the port connected to Arduino—in this example, the port is /dev/tty.usbmodem1411, so I replaced the word REPLACE with 5.

Now for the draw loop. It comprises only two lines. The first line draws the background every time it goes through the loop, which basically clears the slate for each layer. The second line draws a circle with the ellipse() function. The ellipse() function takes four arguments; the x-position of the center of the circle, the y-position of the center of the circle, the width of the circle, and the height of the circle.

We're drawing the circle at the dead center of the sketch, at 150 pixels from the left side and 150 pixels from the right side. For the width and height of the circle, we're reading the 0 analog pin from the Arduino. Here's the code to do that:

```
void draw() {
  background(255);

  ellipse(150, 150, arduino.analogRead(0), arduino.
analogRead(0));
}
```

If everything is working properly, when you move your hand in front of the sensor, the circle should grow larger; when you move your hand away from the sensor, the circle should become smaller.

More Prototyping Tools

Processing and Arduino are the most ubiquitous prototyping tools for this kind of work, but they exist in a much larger ecosystem that has spurred even more specific and exciting tools. Here are a few of my favorites.

ADAFRUIT FLORA

FLORA is a circular Arduino board by Adafruit (Figure 9-20) that's made specifically for wearables projects. You can buy special conductive thread and sew the board into your clothes (along with specially-designed FLORA sensors and actuators). This is a really useful Arduino to have.

FIGURE 9-20
The FLORA platform from Adafruit

INTEL CURIE

The Intel Curie chip (Figure 9-21) powers the Arduino 101 board, which was made specifically to prototype wearable devices. It has built-in Bluetooth LE connectivity as well as a six-axis accelerometer/gyroscope. What makes this board even more powerful are Intel's IQ Software Kits, which include out-of-the-box abilities like identity verification, notifications, step/calorie tracking, data visualizations, and social media integrations. The Curie chip already powers a few commercial smart watches, such as the TAG Heuer Connected watch.

FIGURE 9-21
Intel's Curie board

PARTICLE

Particle makes Arduino-like microcontroller boards that are focused on connectivity. There are two main Particle boards (Figure 9-22): the smaller Photon, which has built-in WiFi, and the larger Electron, which has built-in 3G connectivity and comes with a SIM card. The on-board connectivity isn't what makes these boards special, though; tons of Arduinos come with on-board radio antennas, but what makes Particle stand out is its cloud integration.

Particle makes it easy to connect your device to the internet, but, more important, Particle makes it easy for people who don't know how to set up servers to log events and connect their projects to other services, such as If This Then That (IFTTT).

FIGURE 9-22
Particle boards

ANDROID MODE IN PROCESSING

One of the coolest things that Processing can do is export your sketches directly to Android phones! It's extremely easy to set up, and very few changes are necessary to make your normal Processing sketches Android-ready. Android Mode (Figure 9-23) gets a lot more useful for prototyping when you think about the sensors and actuators that are on an Android phone. Generally, a (relatively) cheap Android phone has built-in WiFi, Bluetooth, cellular connectivity, an accelerometer, a gyroscope, a camera, a microphone, a large amount of local storage, and a big touch screen. It's basically an extremely capable prototyping platform that's ready to go without plugging anything in to it.

FIGURE 9-23
Bluetooth sniffing app made in processing in Android Mode

Android–Ketai

Ketai is an extremely useful Processing library that's built for Android Mode. Ketai gives you access to everything on the phone, such as the accelerometer, gyroscope, light sensor, camera, and microphone; but it also has some other useful things like data storage, computer vision (face detection), location information, and internet connectivity.

Android Watch

Android watch, in this instance, is not talking about Android Wear, but when trying to prototype a smartwatch, I found out that you can actually find watches that run full Android. I can't really think of why someone would want a watch that runs Android, but they definitely work really well with Processing Android Mode.

Pulling It All Together

I understand that this technical stuff can be pretty intimidating for someone who doesn't have any technical experience, but one of the major reasons the wearable community and many other spaces that involve hardware are thriving right now is because of the open source community. All of this stuff was completely out of nontechnical people's reach a decade ago, but there are millions of free, open source resources for people of every level of technical understanding. I know it's scary, but you can definitely do it, and when you start getting things working, it's absolutely thrilling and you'll be hooked. Just give it a shot!

[10]

Selling the Invisible

One of the most difficult aspects about the wearables business is
the business part. You must convince someone to buy an object
that might not be all that interesting or sophisticated in itself,
but is a crucial touchpoint that contributes to an ongoing service.
In most cases, the *device* is completely decoupled from the *value*
of the service, and if you're doing anything new, it's even more
difficult. Selling wearables is a complicated balance of storytelling, finding your niche, and figuring out the right way to make
money from a service. This chapter covers what I've learned from
interviewing and actively working with wearables startups and
resellers as well as a couple theories about how people might
think about objects and how the design of these objects can have
an impact on sales.

Convincing people of the value of a service is a difficult thing,
even in purely screen-based services. People inherently respond
to things that they can actually see. For products like smartwatches, the cool stuff is right there on your wrist, and what
you're getting out of it is more obvious. However, current smartwatches aren't really all that great in terms of providing a service;
they pretty much just show you your phone notifications on your
wrist instead of your phone. With devices that are more oriented
around providing an ongoing service, it's significantly more difficult to demonstrate value.

Hammers and Memories

For an example, let's look at a hammer. Hammers have a long handle;
some even have little ridges for your fingers; and at the head, most all
hammers have a heavier metallic flat front and a curved back. Even if
you didn't know what a hammer was, you could pick it up and quickly

figure out that you're supposed to hit stuff with it. Now let's look at the Narrative Clip, which you can see in Figure 10-1. As previously described in Chapter 5, it's a small square that you wear on your shirt. If I didn't have any context or information about it, I'd never know that the thing was a camera that unobtrusively took a picture every 30 seconds. Now, it doesn't really matter if you can immediately understand what a thing does when you look at it, but for our entire history of using tools, until very recently, they were physically designed around the task for which they were meant to perform in a more obvious way. The difference today is that electronic components are so small that those physical design constraints, cues, and attributes don't really exist, and this makes it more difficult to explain the functionality of the tool. Because the object requires a context for value to be realized, we must tell a story around these objects.

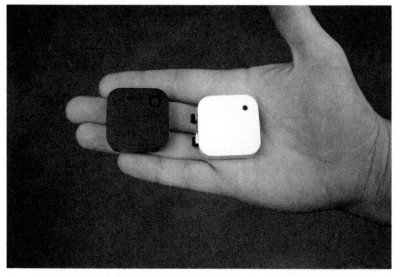

FIGURE 10-1
The Narrative Clip 2 and the original Narrative Clip

SERVICE AVATARS AND STORY TELLING

Beyond the basic understanding of functionality, a more important disconnect exists between the tool and the value that the tool provides. With a hammer, the value is right there in the physical product: you hit something and it does its job. But that's not the case with service-oriented wearable devices. In Chapter 7, we talked about Mike

Kuniavsky's concept of wearable devices essentially being avatars for the service that they enable, and the value of the device coming from the service rather than the device itself. This disconnect presents a big problem when you're trying to explain the value of the service.

Narrative is a great example of this. The Narrative Clip itself is incredibly important to the service because it takes the pictures that enable the larger service that Narrative provides: memories. I tried the Narrative clip because I was interested in the device and what it represented, but I had absolutely no idea how the Narrative service would change my understanding of my own memory. Scrolling through the first month of images showed me things that I might have known somewhere in my head but choose to ignore, such as how much time I actually spend with my work as opposed to my family, or how much time I spend on airplanes as opposed to the destination. I never really thought about time that much, but when it's externalized in such a clear and honest way, you can't ignore the reality of the situation. Not only that, but in reviewing the pictures it became abundantly clear what was most important to me. I saw these seemingly insignificant, everyday moments with my wife, Marita, that I had already forgotten, but the pictures brought them back and helped me to recognize how special these moments are.

FIGURE 10-2
Dinner with Marita

One of the most consistent things I emphasize when working with wearables startups is the need to shift the emphasis from the device features to the end result of those features. The Narrative Clip took these pictures, and as a device, it did its job as expected, but in no way did its list of features convey the value that Narrative's service delivered. Granted, there's absolutely no way that Narrative could have effectively explained to me what I'd get out of its product before using it, but placing more emphasis on the moments that the camera captures rather than its 8-megapixel image quality is a really great start. It makes sense that the Narrative Twitter account posts almost exclusively pictures that users took with the clip, and its website is far more focused on the photographs from the Narrative community than championing the camera.

Retail

Telling the story of a new service isn't just difficult for the company making the devices, it's also a headache for the people working at the stores that sell the products. Most fitness trackers don't look all that different from one another, smartwatches are either round or an Apple Watch, anything that isn't a fitness tracker or a smartwatch just looks like an anonymous minimalist mystery box when you're walking down the electronics aisle. Even if you went in the store with some kind of awareness of the physical object in that aisle, you couldn't possibly know how awesome it is until you try it.

Let's take the Pebble Core (Figure 10-3) as an example. The Pebble Core is described by Pebble as an "ultra wearable," which sounds cool, but doesn't give you any kind of indication of what the device does. Digging in a little bit on the company's Kickstarter campaign reveals that Core is a small Android-powered minicomputer that has a button on it that you can program to do different things. You can even stream music with it away from your phone because it has a 3G chip and is integrated with Amazon's Alexa personal assistant somehow. That sounds pretty awesome to me. I bought one as soon as I learned of its existence. However, if I had to justify this purchase to my wife, she's going to think I'm just wasting my money buying more electric junk. I cannot imagine an eventuality in which she would be walking down an aisle in a retail store, see this device, think it's pretty cool, and decide to buy it. Yet, I know for a fact that about a week after it comes to our house, she's going to either steal it or have me get one for her, and probably another one for our dog.

FIGURE 10-3
The Pebble Core

TARGET'S OPEN HOUSE

In most retail environments, wearables and other internet connected products are shown in brand-supplied display cases, emphasizing sexy industrial design to get your attention guide you to a list of features that might encourage a purchase. Those features might indeed get your foot in the door, but they're a lot more convincing when you can see them in action. This is where retailing giant Target is pushing the boundaries with its *Open House*. Open House is Target's public retail laboratory in downtown San Francisco. It was built from the ground up to tell the story of wearables and smart home products that might be tough to truly understand without seeing them in action.

As it exists now, the space is the third iteration of the retail lab, dedicated to bridge the gap between new technology and consumers. Patricia Adler, experience manager at Target SF, explains:

> In our initial research talking with startups and makers, we found they were a little frustrated that their products weren't fully understood by relying on packaging design. So we decided to take the products off the shelf and out of their boxes. We believe showing the products *in situ* helps people better understand how these products actually work in their homes and can create solutions for their everyday lives.

The first thing you see when you walk in is a translucent wireframe of a home that lets you run different scenarios in order to see how these products work in context. For example, a scenario in the bedroom section (Figure 10-4) is the "health check" that walks you through the data you get from your Jawbone UP and an internet-connected scale which is displayed on the rainbow projector wall. After walking through the home, you run in to a hands-on demonstration area where you can touch and play with the devices and learn more information about features on large touch-screen tables.

FIGURE 10-4
The bedroom of the Open House

Although it's a pretty specific case, this is the only real attempt I've seen to demonstrate time-based services in a retail environment (Figure 10-5). The space definitely does a great job of showing off the devices, but far more emphasis is placed on demonstrating the devices in context and really showcasing the true benefits of the services they enable.

FIGURE 10-5
Trying out new devices at Target's Open House

Pulling It All Together

One-off experiments that you do in your spare time are fun, but we can do a lot more with these devices and the data that they collect when there are millions of them. To do that, we must sell them. Due to the time-based nature of these services and the lack of comparable services, they're really difficult to sell via traditional channels. There's a lot of people working on this, but it all really comes down to communicating value by giving context to the story, either by direct demonstration or by a narrative.

Moving Forward

We Can

I started this book by describing much of the technology industry as surfing a wave propelled by the idea of *because we can*, meaning that a lot of things put out in the world originate from a technology or business-oriented perspective, as opposed to human-oriented; we're just making stuff because we can, without really thinking it through. You know these products when you see them buried in the back of the junk drawer in your kitchen or exiled up on a closet shelf somewhere. They're usually the result of some idea that *seemed* interesting but was never really tested, or maybe it was a good concept that was poorly executed. These products are always going to be out there—There will always be someone who will buy a bracelet that zaps you in order to break a bad habit, or a bulky wearable that lets you play back the last 60 seconds of audio. There's nothing we can do about that. The more dangerous products are those that are actually popular and introduce behaviors and habits that are negative for the user.

I'm not saying that all smartwatches are bad and provide no value—a lot of them are decent fitness trackers, and there's some great location-based stuff happening in Androidwear—but for the most part, they just take the notifications from your phone and put them on your wrist. This doesn't seem *terrible*, per se, but when we step back and look at our overall relationship with technology, we need to be honest with ourselves and think about what we *really* want. Are we better off with a device that primarily functions as a reminder to look at our phones? And, does that give us more or less control in our lives?

We can do better. We've must make some tough decisions about the products and services we create. We cannot base these decisions on whether we think we can convince people to buy them, or base them on what's out there right now. What we currently understand about wearable technology is going to be very different very soon, and we've got to move forward with the understanding that these things are going to change people's lives for good or for bad; thus, we must do our best to move forward doing everything we can to improve people's lives.

Still in the Wild West

I say that we're still in the Wild West because even for as far as we've come, we still have a lot to learn about wearable tech. The past few years alone have introduced some really great technology, and I guarantee that in another five years, the wearables landscape will be completely different, and unrecognizable compared to today. Thanks to advances processing power, mobile communications networks, machine learning, batteries, and manufacturing, we're about to experience a very significant shift in how we understand our relationship with computers as a whole. These are the things I'm most excited about.

MIXED REALITY

Mixed reality (MR) is finally getting to the point that it's going to be both portable and useful. I am thoroughly convinced that MR systems will replace desktop computers, laptops, and televisions. MR will enable us to connect, work, and have fun with one another in entirely new and exciting ways.

FIGURE 11-1
Mixed-reality concept from Magic Leap

COGNITIVE WEARABLES

Cognitive wearables are still in their early phase, but will become much more accurate and powerful, allowing us to externalize basic things like stress and relaxation, but also more nuanced parts of our mental lives, such as emotions. Having access to this information will give us the ability to know much more about ourselves, and empower us to take more control over our lives.

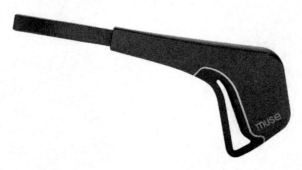

FIGURE 11-2
Muse EEG headband

MACHINE LEARNING

Machine learning technology is rapidly expanding, and as activity trackers, camera-based technologies, cognitive wearables, and other devices collect more detailed information about our lives, machine learning will bring that information to us in a way that will make it truly useful.

FIGURE 11-3
An illustration of reinforcement learning exploration

SENSORY AUGMENTATION

Sensory augmentation will not only give us more control over the raw information that we currently perceive, it will give us the ability to sense things outside of our natural abilities and explore the world in new ways.

FIGURE 11-4
Woman wearing sensory augmentation earbuds by Here One

Go Forth!

I cannot think of a better time to be in the wearables industry. Right now, more than any other time in history, we as designers and technologists have the tools and skills to create things that solve real problems. We're beginning to get an idea of what works and what doesn't, we're understanding the human mind better than we ever have, we have access to the tools we need to create new things, and we have the methodologies to determine whether those work. These emerging technologies present a massive opportunity for all of us to grow as a civilization, and whether we like it or not, we're going to have a hand in that change. The bottom line is that the more knowledge we have of ourselves, our immediate context, and our greater context, the more control we have over our lives, and the more control we have over our lives, the better our lives will be.

[Index]

Learn from experts.
Find the answers you need.

Sign up for a **10-day free trial** to get **unlimited access** to all of the content on Safari, including Learning Paths, interactive tutorials, and curated playlists that draw from thousands of ebooks and training videos on a wide range of topics, including data, design, DevOps, management, business—and much more.

Start your free trial at:
oreilly.com/safari

(No credit card required.)

2017
"|18. 3x lad 7/18

CPSIA information can be obtained
at www.ICGtesting.com
Printed in the USA
BVOW11s0752161216
470954BV00002B/2/P

9 781491 944158

O'REILLY®

Designing for We

Now may be the perfect time to enter the wearables industry. With the range of products that have appeared in recent years, you can determine which ideas resonate with users and which don't before leaping into the market. In this practical guide, author Scott Sullivan examines the current wearables ecosystem and then demonstrates the impact that *service design*, in particular, will have on these types of devices going forward.

You'll learn about the history and influence of activity trackers, smartwatches, wearable cameras, the controversial Googl   r devices that have co   t period. This book also dives into many other aspects of wearables design, including tools for creating new products and methodologies for measuring their usefulness.

You'll explore:

- Emerging types of wearable technologies
- How to design services around wearable devices
- Key concepts that govern service design
- Prototyping processes and tools such as Arduino and Processing
- The importance of storytelling for introducing new wearables
- How wearables will change our relationship with computers

"An excellent guide on how different forms of wearables developed, specific examples of what is happening now and how it works, and how people can practically frame their own ideas for product development. Use it to understand the different types of hardware and the user needs that are being met now and could be met in the future."

—Alastair Somerville
Sensory Design Consultant
at Acuity Design

Scott Sullivan is an applied designer at Adaptive Path, where he designs and develops services in complex technological ecosystems. Scott also works with wearable startups to develop stories and services to make the devices meaningful.

USER EXPERIENCE

US $29.99   CAN $34.99
ISBN: 978-1-491-94415-8

52999

Twitter: @oreillym
facebook.com/orei

W9-AIM-385